Systematic Evidence Review
Number 25

Screening for Cervical Cancer

Prepared for:
Agency for Healthcare Research and Quality
U.S. Department of Health and Human Services
2101 East Jefferson Street
Rockville, MD 20852
http://www.ahrq.gov

Contract No. 290-97-0011
Task No. 3
Technical Support of the U.S. Preventive Services Task Force

Prepared by:
Research Triangle Institute/University of North Carolina
3040 Cornwallis Road
PO Box 12194
Research Triangle Park, NC 27709

Katherine E. Hartmann, MD, PhD
Susan A. Hall, MS
Kavita Nanda, MD, MHS
John F. Boggess, MD
Dennis Zolnoun, MD

January 2002

Preface

The Agency for Healthcare Research and Quality (AHRQ) sponsors the development of Systematic Evidence Reviews (SERs) through its Evidence-based Practice Program. With guidance from the third U.S. Preventive Services Task Force[*] (USPSTF) and input from Federal partners and primary care specialty societies, two Evidence-based Practice Centers—one at the Oregon Health Sciences University and the other at Research Triangle Institute-University of North Carolina—systematically review the evidence of the effectiveness of a wide range of clinical preventive services, including screening, counseling, immunizations, and chemoprevention, in the primary care setting. The SERs—comprehensive reviews of the scientific evidence on the effectiveness of particular clinical preventive services—serve as the foundation for the recommendations of the third USPSTF, which provide age- and risk-factor-specific recommendations for the delivery of these services in the primary care setting. Details of the process of identifying and evaluating relevant scientific evidence are described in the "Methods" section of each SER.

The SERs document the evidence regarding the benefits, limitations, and cost-effectiveness of a broad range of clinical preventive services and will help to further awareness, delivery, and coverage of preventive care as an integral part of quality primary health care.

AHRQ also disseminates the SERs on the AHRQ Web site (http://www.ahrq.gov/uspstfix.htm) and disseminates summaries of the evidence (summaries of the SERs) and recommendations of the third USPSTF in print and on the Web. These are available through the AHRQ Web site (http://www.ahrgq.gov/uspstfix.htm), through the National Guideline Clearinghouse (http://www.ncg.gov), and in print through the AHRQ Publications Clearinghouse (1-800-358-9295).

We welcome written comments on this SER. Comments may be sent to: Director, Center for Practice and Technology Assessment, Agency for Healthcare Research and Quality, 6010 Executive Blvd., Suite 300, Rockville, MD 20852.

Carolyn Clancy, M.D.
Acting Director
Agency for Healthcare Reseach and Quality

Robert Graham, M.D.
Director, Center for Practice and
Technology Assessment
Agency for Healthcare Research and Quality

---

[*]The USPSTF is an independent panel of experts in primary care and prevention first convened by the U.S. Public Health Service in 1984. The USPSTF systematically reviews the evidence on the effectiveness of providing clinical preventive services--including screening, counseling, immunization, and chemoprevention--in the primary care setting. AHRQ convened the third USPSTF in November 1998 to update existing Task Force recommendations and to address new topics.

The authors of this report are responsible for its content. Statements in the report should not be construed as endorsement by the Agency for Healthcare Research and Quality or the U.S. Department of Health and Human Services of a particular drug, device, test, treatment, or other clinical service.

Table of Contents

List of Tables and Figures viii

Abstract ix

I.   Introduction 1
       Background 1
       Uniform Terminology for Cervical Lesions 2
       Burden of Suffering 2
       Epidemiology 3
            Risk Factors 3
            Role of Human Papilloma Virus 3
            Screening Failures 5
       Screening Tools 6
            Cervical Cytology: Conventional and New Technologies 7
            Tools for HPV Testing 9
       Health Care Interventions 10
       Prior Recommendations 12
            US Preventive Services Task Force 12
            Recommendations of Other Groups 13
       Analytic Framework and Key Questions 14
            Analytic Framework 14
            Key Questions 15
            Relevant Outcomes 16

II.   Methods 26
       Literature Search Strategy 26
       Inclusion/Exclusion Criteria 27
       Literature Reviewed 28
            Citation Database 28
       Screening of Articles 28
            Key Question 1: Screening Among Older Women and After Hysterectomy 29
            Key Question 2: New Technologies for Cytology 30
            Key Question 3: What is the Role of HPV Testing? 32
            Cost and Harms 32
       Literature Synthesis and Preparation of Systematic Evidence Review 34
            Data Abstraction and Development of Evidence Tables 34
            Peer Review Process 35

III.   Results 39
       Key Question 1 39
            Screening Among Women Age 65 and Older 39
            Incidence and Age 41
            Prevalence 42

Table of Contents

        Screening History and Interval 43
        Screening among Women Who Have Had a Hysterectomy 44
        Summary 45
    Key Question 2 46
        New Methods for Preparing or Evaluating Cervical Cytology 46
        Neural-network Rescreening 50
        Computerized Rescreening 53
        Harms 54
        Benefits 54
        Costs 55
        Summary 57
    Key Question 3: Role of HPV Testing in Cervical Cancer Screening 58
        Screening Use of HPV Testing 59
        HPV Screening for Detection of High-grade Cervical Changes 60
        HPV Screening for Detection of Low-Grade Cervical Changes 61
    Triage USE of HPV Testing 62
        Benefits 65
        Risks 66
        Costs 66

IV.   Discussion 75
    Context 75
    Major Findings and Limitations of the Literature 76
        Who Should be Screened and How Often? 76
    New Methods for Preparing or Evaluating Cervical Cytology 78
    The Role of HPV Testing in Cervical Cancer Screening and Triage 79
    Benefits and Harms 80
    Future Research Needs 81

References 83

List of Appendices

Appendix A.   Acknowledgments A-1
Appendix B.   Methods B-1
Appendix C.   Evidence Tables C-1

List of Tables

Table 1.	Technical Characteristics of HPV Testing Methods	21
Table 2.	Recommendations of Other Groups about Pap Smear for Cervical Cancer Screening	22
Table 3.	Overall Inclusion and Exclusion Criteria	36
Table 4.	Literature Search Results (1995-2000)	37
Table 5.	Disposition of Articles Identified by Literature Search	38
Table 6.	Performance of ThinPrep® in a Prospective Cohort	69
Table 7.	Studies with Screening Uses of HPV Testing	70
Table 8.	Performance of Screening HPV Testing for Detection of High-grade Abnormalities	71
Table 9.	Performance of Screening HPV Testing for Detection of Low-grade or More Severe Abnormalities	72
Table 10.	HPV Testing Among Women with Abnormal Pap test Results	73
Table 11.	HPV Testing as a Triage Tool Among Women with an Abnormal Pap Test for Detection of HSIL	74

List of Figures

Figure 1.	Map of Cervical Cytology Classification Schemes	19
Figure 2.	Screening for Cervical Cancer: Analytic Framework	20

Structured Abstract

Context: Methods that improve detection of serious cervical lesions while minimizing excess screening are the key to advancing cervical cancer prevention.

Objective: To examine the evidence about benefits and harms of screening among older women (ages 65 and older) and those who have had hysterectomies, and to examine the diagnostic performance of new technologies and human papilloma virus (HPV) testing for detecting cervical lesions.

Data Sources: We identified English-language articles on cervical neoplasia, cervical dysplasia, and screening from a comprehensive search of the MEDLINE database from 1995 through June 2000. In addition, we used published systematic reviews, the second *Guide to Clinical Preventive Services*, and peer review to assure a complete update of specific topics.

Study Selection: We included articles that reported on screening for squamous cell carcinoma of the cervix if they included the age distribution of the study population and presented analyses stratified by age or if they included hysterectomy status as a covariate. For diagnostic tools, we required that the test be used as part of a screening strategy, that the method be compared with a reference standard, and that all cells of a 2x2 table can be completed.

Data Extraction: We extracted the following data from articles addressing screening among older women and those who have had a hysterectomy: study design, objectives, location and timeframe, source of the data (e.g., population-based registry), participants, screening program used, outcomes and measures, and results relevant to age and screening interval. For articles about diagnostic tests, we extracted study design, test methods, location, patient population, outcome measures (emphasizing documentation of the reference standard), prevalence of lesions, and test characteristics including sensitivity, specificity, and predictive

Structured Abstract

values. We used scoring checklists to summarize strengths of the publications; we also evaluated the validity of each article and the overall quality of the evidence.

Data Synthesis: The evidence about age and hysterectomy is observational, predominantly from population- or care-based data. The findings are consistent: risk of cervical cancer or abnormalities falls with age; high-grade and more severe lesions are detected in fewer than 1 per 1,000 Pap tests among women older than 60 who have had prior screening; and longer histories of prior normal Pap tests further reduces risk. After hysterectomy, high-grade vaginal lesions are rare, fewer than 2 to 4 per 10,000 tests. The literature about new diagnostic tools is limited by lack of histologically validated performance. Using tools such as liquid cytology, neural-net rescreening, and computer-based review algorithms improves sensitivity; however, this improvement is predominantly for detection of low-grade lesions. The impact on specificity is poorly documented. Sensitivity of HPV testing for screening detection of high-grade lesions is competitive with conventional cytology (roughly 82%); specificity is lower (78%); and negative predictive value is good (99%). For triage of women with abnormal Pap tests, sensitivity for detecting high-grade lesions is 85%, specificity is 60%, and negative predictive value is 97%.

Conclusion: The yield of screening among older women who have been previously screened decreases with age; if recommendations are not modified, older women are disproportionately likely to have evaluations for false-positive findings. The prior recommendation of the US Preventive Services Task Force to discontinue Pap testing after hysterectomy for benign disease is supported. For making decisions about screening modality in US populations, evidence about these new technologies for cytology screening and HPV testing is currently limited. Controlled trials and prospective cost evaluation of new screening strategies

Structured Abstract

in each of these areas are required. Important trials will be completed in 2001 that may clarify our conclusions.

Chapter I. Introduction

I. Introduction

Background

Since introduction of cytologic screening for cervical cancer using the Papanicolaou (Pap) test in the 1950s, the incidence of invasive cervical cancer in the United States has fallen more than 100%.[1] No other cancer screening program has been more successful. This fall occurred despite an increase in risk factors for cervical cancer, such as younger age at initiation of sexual intercourse, more sexual partners in a lifetime, and greater prevalence of human papilloma virus (HPV) infection and cigarette smoking.

Success in prevention reflects three factors: (1) progression from early cellular abnormalities, termed low-grade dysplasia, through more severe dysplasia, to carcinoma *in situ* and invasive cancer is generally slow, allowing time for detection; (2) associated cellular abnormalities can be identified; and (3) effective treatment is available for premalignant lesions. Consequently, invasive squamous cell carcinoma of the uterine cervix is a highly preventable disease.

Introduction of screening programs in populations naïve to screening reduces cervical cancer rates by 60% to 90% within three years of implementation.[2,3] This reduction of mortality and morbidity with introduction of screening with the Pap test is consistent and dramatic across populations. As a result, Pap testing is one of the few preventive interventions that has received an "A" recommendation from the US Preventive Services Task Force (USPSTF) in the absence of randomized trials demonstrating effectiveness.[4]

Chapter I. Introduction

Uniform Terminology for Cervical Lesions

Figure 1 depicts the relationship between varied systems of nomenclature for describing cytologic and histologic findings. In this report we use terminology from the Bethesda System — low-grade squamous intraepithelial lesion (LSIL) and high-grade squamous intraepithelial lesion (HSIL) to describe cytology findings. When possible, we use the cervical intraepithelial neoplasia (CIN)—specifically, categories CIN 1-3—to refer to histologic findings. As necessary, we present histology findings for the groupings LSIL or HSIL if this is the most detailed summary of results provided by the investigators.

Burden of Suffering

In the United States, approximately 12,800 new cases of cervical cancer are diagnosed and 4,800 deaths occur each year.[5] Incidence of cervical cancer is decreasing; US rates have decreased from 14.2 new cases per 100,000 women in 1973 to 7.8 per 100,000 in 1994. For each woman with invasive disease, there will be 4 with carcinoma in situ and 10 with cervical dysplasia.[6] Despite falling incidence, cervical cancer remains the ninth most common cause of cancer deaths.[5] Of the cancer prevention goals established in *Healthy People 2000*, including colorectal, lung, and breast cancer, cervical cancer mortality rates were the furthest off target at the mid-course review. The target for cervical cancer was reduction of mortality to 1.3 deaths per 100,000 women; the current rate remains near 2.7 deaths per 100,000, down only slightly from 2.8 per 100,000 in 1987.[7]

Detection of cervical cancer in its earliest stages is lifesaving, as survival of cancer of the cervix uteri depends heavily on stage at diagnosis. Although 91.5% of women will survive 5 years when the cancer is localized, only 12.6% will survive distant disease.[1]

Chapter I. Introduction

Epidemiology

Risk Factors

Squamous cell carcinoma of the cervix and its cytologic precursors are conditions of sexually active women. Infection with high-risk strains of HPV, generally acquired sexually, is the most important risk factor for cervical cancer. The role of HPV is described in greater detail below.

Risk factors relating to sexual behavior that are associated with increased risk include onset of intercourse at an early age and a greater number of lifetime sexual partners. These behavioral risks appear to persist even after controlling for effects of HPV infection.[8] A higher number of lifetime sexual partners in the male partners of cervical cancer cases compared to controls has also been noted in case-control studies.[8] Occupations are related to cervical cancer only through relationships with sexual behavior.[8] Cigarette smoking is the only nonsexual behavior consistently and strongly correlated with cervical dysplasia and cancer, independently increasing risk two- to fourfold.[9-11] In the United States, black race and low socioeconomic status are associated with increased risk. Recently, attention has been drawn to a positive family history as a plausible risk factor.[12]

Role of Human Papilloma Virus

HPV plays a central role in the development of cervical cancer. Using modern HPV detection methods, 95% to 100% of squamous cell cervical cancer and 75% to 95% of high-grade CIN lesions have detectable HPV DNA.[13-15]

HPV is a double-stranded DNA virus. The virus is transmitted to the cervix and vaginal tissues primarily by sexual intercourse.[16,17] HPV can infect and persist in vulvar, vaginal, and cervical tissue throughout a lifetime. This family of viruses includes those responsible for

Chapter I. Introduction

genital condylomata or warts, squamous cell carcinomas of the genital tract including vaginal and vulvar cancers, and cervical cancer. More than 70 strains or types of HPV have been classified. For instance, HPV Types 6 and 11 cause warts; other types have oncogenic properties.

The best characterized types associated with cervical cancer are Types 16 and 18.[18] These are officially recognized by the International Agency for Research on Cancer (IARC) and the World Health Organization (WHO) as carcinogenic infectious agents. The primary difference between "oncogenic" and "non-oncogenic" virus is the interaction of two viral genes, E6 and E7, that influence cell cycle control mechanisms. Oncogenic E6 and E7 gene products can "cripple" a normal cell's ability to control cell proliferation, which in some instances leads to cancer.

HPV is a necessary but not sufficient precursor of squamous cell carcinoma of the cervix. Among women without cervical cytology abnormalities at baseline, those with high-risk HPV types have a relative risk of developing high-grade cervical lesions (CIN 2 - 3, CIS) that is 58- to 71-fold higher than the risk for those without detectable HPV.[19,20]

The natural history of HPV acquisition, clearance, persistence, and possible re-infection is complex. To promote cervical cancer abnormalities, the virus must become integrated into the host genomic DNA. This event, which is essential for cancer progression, appears to be rare. In the absence of viral integration, the normal viral lifecycle produces morphologic changes in the cervical epithelium characteristic of low-grade dysplasia (LSIL). With viral integration, the oncogenic effect of the E6 and E7 proteins is enhanced and cellular changes characteristic of high-grade dysplasia and ultimately cancer are observed.[21,22]

Chapter I. Introduction

Inter-related host factors such as age, nutritional status, immune function, smoking, and possibly silent genetic polymorphisms modulate incorporation of viral DNA. Studies of the time required from infection to incorporation are challenging to interpret because assays for viral DNA integration are difficult to perform.[18] Taken as a whole, however, nearly 100% of cases of carcinoma in situ and cancer are estimated to have integrated HPV DNA compared to a small minority of low-grade lesions.[15]

The transition time from simple viral infection to integration of DNA is unknown and may be influenced by the risk profile of population studied. For instance, although the prevalence of HPV infection is higher among immunocompromised hosts such as HIV-infected women, the speed of progression to cervical cancer is not increased. Natural history studies confirm that in the vast majority of cases, the course of infection and cervical abnormalities that progress do so in an orderly fashion from less severe to more severe lesions; de novo HSIL with HPV incorporation appearing in a short interval is rare. Thus, the sequence associated with HPV infection and development of cervical cancer is as amenable to surveillance as are cytologic changes.

In the United States, peak incidence and prevalence of HPV infection occur among women under age 25.[23] More than 30% of postmenopausal women, however, have detectable HPV DNA using polymerase chain reaction (PCR) detection methods.

Screening Failures

In the United States, incident cases of squamous cell carcinoma can be attributed to different categories of failures of screening. Between 50% and 70% of cancer cases occur among women who have never been screened or who have not been screened within the past 5 years.[24,25] Among women who have been screened, failures may occur in 3 ways. First,

Chapter I. Introduction

abnormalities may be identified by screening but not properly treated as a result of patient or provider failure to follow-up the abnormalities; this may occur in approximately 22% to 63% of those who receive proper screening. Second, serious abnormalities may not be present at the time of screening, and progression occurs between recommended screening intervals. When the interval is annual, such progression is rare; with 3-year intervals, this may happen in up to 50% of diagnosed cases.[26] Progression is more common among women under 45 years of age. Third, abnormalities may be present but are not detected by the screening test (approximately 14% to 33%).[24,25] The last category of failures, those related to the screening test, can be further subdivided into those that represent sampling error (cells from the abnormal area were not obtained and so could not be identified in the specimen) and those that reflect detection error (the abnormal cells are included in the specimen and are not identified as abnormal).

Screening Tools

This report is focused on cervical cancer screening tools for clinical use in primary care settings. We consider traditional cytology and new cytologic technologies that are currently available to practitioners. HPV testing to identify specific types of HPV is less broadly available. Some HPV test methodologies such as Southern blot are appropriate as a gold-standard assay but not practical for widespread implementation in a cervical cancer screening program. As a result, we have focused on HPV tests suitable for high-volume use as part of primary clinical care. Basic descriptions of available cytologic and HPV screening tools are provided below.

Chapter I. Introduction

Cervical Cytology: Conventional and New Technologies

Ordinarily, cervical cancer screening specimens are obtained at the time of pelvic examination during the portion of the examination when a speculum is used to visualize the cervix and obtain a sample for cytology. The goal of sampling for cytology purpose is to sample the transformation zone: that area of the cervix where physiologic transformation from the columnar cells lining the endocervical canal to the squamous cells covering the ectocervix occurs. Cervical dysplasia and cancers arise in the transformation zone. The transformation zone is easily sampled in younger women because it is on the surface of the cervix. With increasing age, however, the transformation zone is more likely to be higher in the endocervical canal.

Various sample collection tools are available to accomplish the goal of sampling both the ectocervix and the endocervix. Meta-analysis of randomized trials supports recommendations for combined use of a spatula, preferably an extended tip spatula, for sampling the ectocervix and a brush for sampling the endocervix.[27]

Conventional cervical cytology specimens are prepared by using the collection tool to smear the specimen onto a glass microscope slide while the woman is in the examination room. Two slides or two distinct areas of the same slide are prepared to represent the ectocervical and endocervical samples. The slides are then immediately sprayed with or placed in fixative. Slides are sent to the cytology laboratory and read by technicians who review the entire slide at 10x magnification, systematically in 2mm sections.

Thin layer cytology is a variation of conventional cytology. The Food and Drug Administration (FDA) has approved two systems: they are ThinPrep® (Cytyc, Boxborough, MA) and AutoCyte PREP® (TriPath Imaging, formerly Autocyte, Burlington, NC), approved in May

Chapter I. Introduction

1996 and June 1999, respectively. Specimens are collected in the same fashion as those for conventional cytology; however, rather than smearing the sample onto slide(s), the sample is suspended in the fixative by stirring the specimen collection spatula and brush in the fixative solution. The container is sealed and the specimen sent to the cytology laboratory in solution rather than on slides; this theoretically improves the probability of transferring a representative sample of cells to the slide. In the laboratory, technicians disperse the sample in the fixative and then collect the cells on a filter and transfer them to a microscope slide in a monolayer. Immediate fixation and uniform spread of the cells are designed to reduce detection errors by assuring that cells are well preserved, not obscured, and more easily assessed by cytology technicians.

Both conventionally prepared and thin layer specimens read by cytotechnologists are subject to random manual rescreening of slides that were interpreted as normal at a minimum rescreening rate of 10%, as required by the Clinical Laboratories Improvement Amendments (CLIA) of 1988 (Final Rule, Federal Register, 1992).

Computerized rescreening is designed to automate rescreening of Pap smears initially read as negative by a cytotechnologist. PapNet® (TriPath Imaging, Inc.) uses neural-network technology to interpret computerized images of the Pap slide. The system, approved by the FDA for rescreening use in December 1995, identifies cells or other material on negative Pap slides that require review and creates a summary display of up to 128 images that may contain abnormalities. A cytotechnologist then reviews the summary images and can also return to the original slide using light microscopy. Several countries allow use of PapNet as a primary screening as well as a rescreening technology, and the literature contains studies of both forms of use.

Chapter I. Introduction

Algorithm-based rescreening identifies slides that exceed a selected probability for containing abnormal cells. The AutoPap® 300QC system (Neopath, Inc.—when reviewed, now TriPath), FDA approved in May 1998, can be set for different thresholds that result in 10%, 15%, or 20% review rates. Because a classification algorithm is applied, the slides selected for review are more likely to contain abnormalities than are negative Pap slides randomly selected for review. The system is reported to identify 70% to 80% of those slides misdiagnosed as normal (i.e., false negatives) at the time of manual screening. AutoPap is also employed for both routine screening and rescreening purposes outside the United States, and publications evaluate both types of use.

We excluded review of the performance of AutoSCREEN, a product for interactive, computer-assisted screening and rescreening. Its manufacturers no longer produce it after a corporate merger between the AutoPap parent company and the original producer. We also have not reviewed approaches that require additional time, equipment, expertise, or materials at the time of the clinical pelvic examination, such as speculoscopy, screening colposcopy, or cervicography to photograph the cervix. These are not suitable for widespread implementation in the United States.

Tools for HPV Testing

Cellular changes associated with HPV infection can be seen on visual examination of cervical cytologic or biopsy specimens and, to a lesser degree, at colposcopy. However, the presence of these changes is neither sensitive nor specific.

Direct HPV testing methods for assessing the presence and type of HPV are more promising; they rely on identification of HPV viral DNA. Some versions of testing are qualitative, detecting only the presence or absence of HPV DNA; some are quantitative,

Chapter I. Introduction

estimating the viral burden; and some methods allow assessment of degree of integration of HPV into the host genome. Tests differ in the method for processing the DNA analyzed, the quantity of specimen required, and difficulty of analysis.

Because the family of HPV viruses is large, and because only a small group of those types is associated with cervical dysplasia and cervical cancer, tests that identify specific viral types or panels of high-risk viral types are preferable to detection of all HPV types. A multitude of laboratory methods has been applied to studying HPV. However, these approaches are not all equally applicable to screening. Considering the reliability and relative performance characteristics of the assays in controlled settings in the laboratory and their potential for application to screening, we have adopted the classification of HPV testing technologies presented in Table 1.

This is a modification of the classification used by John Cuzick, his co-authors, and an expert panel of the Health Technology Assessment Board of the UK National Health Service in their *Systematic Review of the Role of Human Papillomavirus Testing Within a Cervical Cancer Screening Program.*[21] Based on consultation within our team, with our USPSTF liaisons, and with other experts, we have focused our review on evaluation of clinically collected cervical samples using Hybrid Capture II (HCII) and on type-specific, the SHARP detection system, or conventional consensus PCR.

Health Care Interventions

Reduction of morbidity and mortality associated with squamous cell carcinoma of the cervix is the ultimate goal of screening. The screening system used must be acceptable to patients and providers and detect abnormalities that are amenable to intervention. The broad

Chapter I. Introduction

spectrum of cervical abnormalities detected by cytology, from low-grade changes to carcinoma in situ, is a particular challenge. HSIL warrants immediate evaluation by colposcopy, biopsy, and endocervical curettage. Patients with CIN 3/CIS receive definitive intervention, such as conization by loop electrosurgical excision procedure (LEEP), laser, or cold knife, to remove the transformation zone and confirm that no invasive disease is present. Patients with CIN 2 may be treated with conization or ablative procedures to remove or destroy the transformation zone. Appropriate intervention for atypical squamous cells of uncertain significance (ASCUS) and LSIL are active areas of research focused on determinants of progression, stability, and regression.

Current clinical care of ASCUS and LSIL increases the vigilance of follow-up and results in colposcopic evaluation for the majority of women with these diagnoses. Evaluation of ASCUS is guided by the cytologic clarification that accompanies the diagnosis: (a) reactive processes are followed with repeat Paps every 6 months until 3 normals; (b) inflammation prompts evaluation for infection, followed by repeat Pap; (c) atrophic changes are treated with topical estrogen, followed by repeat Pap; and (d) ASCUS favoring atypia is evaluated in a fashion comparable to the degree of atypia suggested (i.e., "ASCUS favor LSIL" is evaluated as LSIL). LSIL, in a patient who will be compliant with follow-up and is comfortable with expectant management, can be followed with repeat Paps every 6 months until 3 consecutive negative smears are obtained or progression is noted. Colposcopy and biopsy is preferred for high-risk patients (characteristically broadly defined and not clearly specified) and required for those with immune compromise or prior dysplasia.

As women are screened at younger ages and in larger numbers, and as tests are increasingly sensitive for detecting low-grade changes such as ASCUS and CIN 1, the

Chapter I. Introduction

importance of assessing potential harms of screening becomes clear. The psychological effects of labeling young women with an anxiety-provoking, possibly precancerous condition, and the associated individual, health care system, and societal costs deserve attention. Ideally, screening tools would help guide selection of the intensity of intervention across the spectrum of cervical dysplasia.

For this report, we sought evidence about screening in older age groups and after hysterectomy. We also looked at methods to determine the optimal interval and at the potential contribution of new technologies and HPV testing methods to clinical prevention of cervical cancer. Furthermore, we sought evidence about the cost implications, the population to be screened, and potential harms that follow from screening.

Prior Recommendations

US Preventive Services Task Force

The second edition of the *Guide to Clinical Preventive Services* from the US Preventive Services Task Force gave an "A" to the recommendation of a regular Pap test for all women who are or have been sexually active and who have a cervix.[4]

The Task Force further stipulated (pg 112):[4]

There is little evidence that annual screening achieves better outcomes than screening every 3 years ("B" recommendation). The interval for each patient should be recommended by the physician based on risk factors. There is insufficient evidence to recommend for or against an upper age limit for Pap testing, but recommendations can be made on other grounds to discontinue regular testing after age 65 in women who have had regular previous screening in which the testing has been consistently normal ("C" recommendation). Women who have undergone a hysterectomy in which the cervix was removed do not require Pap testing unless the hysterectomy was performed because of cervical cancer or its precursors.

Chapter I. Introduction

These recommendations were based on limited, although consistent, literature. The use of HPV testing, cervicography, and colposcopy lacked evidence in the literature to support routine use. Recommendations against use were based on other grounds, including poor specificity and costs.

Recommendations of Other Groups

Table 2 summarizes the recommendations of 9 selected US organizations and other international groups and health systems with guidelines based on evidence review. We note information on starting age, interval, adaptation of interval for high-risk women, upper age limit for screening, and discontinuation after hysterectomy.

Intervals range from 2 to 5 years, in some cases with modification for individual risk and in some cases irrespective of risk. The most commonly advised screening interval is 3 years after a specified number of qualifying prior normal smears.

Six of the 9 groups suggest discontinuation among older women with prior normal screening history: The UK National Health Service Cancer Screening Programmes at 64; the American College of Preventive Medicine and the Institute for Clinical Systems Improvement at 65; the Canadian Task Force on Preventive Health Care at 69; and the National Cervical Cancer Screening Programs of Australia and New Zealand at age 70. Three specifically mention discontinuing testing after hysterectomy for benign disease.

No guidelines specifically address the utility or advisability of using new cytology technologies or HPV testing for screening or triage, within a screening system, except in noting that HPV infection is a high-risk factor that may guide choice of interval.

Chapter I. Introduction

Analytic Framework and Key Questions

The RTI-University of North Carolina Evidence-based Practice Center (RTI-UNC EPC), together with members of the USPSTF and other clinical and methodologic experts (see Appendix A), sought to update specific topics in the area of cervical cancer screening that have evolved rapidly in the past 5 years. This systematic evidence review (SER) updates Chapter 9 (pages 105-117) of the second *Guide to Clinical Preventive Services*.[4]

Analytic Framework

Conceptualizing approaches to cervical cancer screening in primary care practice requires an analytic framework. Our framework (Figure 2) is not intended to provide etiologic detail; rather, it depicts the relationship between the progression of disease and the potential points of intervention to prevent morbidity and mortality. These potential points of intervention provide the rationale for the questions undertaken in this systematic review.

The pathway starts with women potentially eligible for screening. We have conceptualized screening as a process that may have more than one component. For example, a woman presenting for care may have her individual risk for cervical cancer assessed based on her past sexual history, medical history, and prior Pap test results. Using that information, clinicians determine the need for a Pap test. If a Pap test is done, the specimen may be prepared by conventional methods or new techniques. The cytology specimen may be read in the conventional manner or with pre-screening or re-screening by computer-assisted methods. Other screening studies, such as HPV testing, may be incorporated before a woman is determined to have normal or abnormal screening results.

Chapter I. Introduction

We have denoted the combination of screening components in Figure 2 the "screening strategy." Key Questions 2 and 3 (see below) focus on the role of specific components within a cervical cancer screening strategy.

In this analytic framework, women with normal findings at completion of a screening strategy return to the routine screening group. Women with abnormal screening tests progress to further evaluation. Evaluation may fail to identify any abnormalities (false-positive screening test results) or result in the diagnosis of cervical dysplasia or cancer. Among those who are diagnosed with an abnormality, the severity of dysplasia or stage of cancer is the primary determinant of treatment options and morbidity and mortality risk. The success of a screening strategy depends on early detection of pathology, which then facilitates early treatment, ultimately resulting in improved length of life, quality of life, or both for women who are screened.

Key Questions

Our key questions are provided below. They appear in Figure 2 as Key Q1, Key Q2, and Key Q3 above the related arrows, which indicate steps in the prevention process and disease progression.

- Key Question 1: Who should be screened for cervical cancer and how often?

In developing the work plan for this SER, we specified this question broadly to prompt discussion of what focus would most contribute to guiding screening in primary care practice. We considered a range of potential topics including age at initiation of screening, need for screening among lesbian women, screening recommendations for women with HIV infection, interval of screening in the general population, screening after hysterectomy, and screening among older women including the relationship between aging and interval. Practical limitations

Chapter I. Introduction

required a narrow focus. Ultimately we concurred that our work would be best focused on the screening needs of older women and on screening after hysterectomy. As a result, we restricted our examination to the evidence about the cervical cancer screening needs of older women who have had a hysterectomy.

Specifically, we asked what are the outcomes (benefits, harms, and costs) associated with screening:

1A. Among women age 65 and older?

1B. Among women who have had a hysterectomy?

Only Key Question 1A is focused on older age, interval among older women, and screening outcomes. *The remainder of the entire SER applies to all women irrespective of age.* For instance, 1B addresses screening among all women who have had a hysterectomy.

- Key Question 2: To what extent do new methods for preparing or evaluating cervical cytology improve diagnostic yield compared to conventional methods? At what cost (harms and economic)?

- Key Question 3: What is the role of HPV testing in cervical cancer screening strategies? Specifically:

    3A. What are the benefits, harms, and costs of using HPV testing as a screening test, or of incorporating HPV testing at the time of the screening Pap test, compared with not testing for HPV?

    3B. What are the benefits, harms, and costs of using HPV testing as part of a screening strategy to determine which women with an abnormal Pap test should receive further evaluation?

Relevant Outcomes

Incidence of cervical cancer, severity of disease at the time of diagnosis, and cervical cancer mortality are relevant to all questions. However, no trials have been conducted to

Chapter I. Introduction

evaluate the effectiveness of cervical cancer screening systems for improving these outcomes. Given the rarity of cervical carcinoma among women who receive any screening, much of this literature examines surrogate endpoints such as the diagnosis of dysplasia by colposcopy and/or cervical biopsies. The outcomes we have been able to assess were determined by the content of the literature. Outcome(s) of interest were tailored to the related key questions in the following fashion.

Outcomes for Key Question 1A and 1B, which address candidates for screening, include diagnosis of dysplasia, severity of dysplasia, progression of dysplasia, diagnosis of carcinoma, stage at diagnosis, and mortality. We retained cost data, if relevant to care in the United States, for background information. We also sought harms related to screening.

For Key Questions 2 and 3, which examine the test characteristics of new methods of preparing and evaluating cervical cytology and the contribution of HPV testing, the relevant outcomes include sensitivity, specificity, positive and negative predictive value, and diagnostic yield when compared to a gold standard. The standard used and the quality of evaluation of normal screening results both vary. Much of this literature is limited by failure to apply the gold standard evaluation to normal subjects, which precludes definitive assessment of specificity. We sought to evaluate the influence of these tools on the diagnoses and features listed above for Key Question 1, and we retained literature relevant to US costs for background.

Chapter II provides an overview of our methods for producing the SER. Chapter III presents the results of our literature search and synthesis organized by the key questions. We discuss the results and the limitations of the literature in Chapter IV with attention to ramifications for future research. Appendix A contains acknowledgments, Appendix B provides

Chapter I. Introduction

additional detail about our methods, and Appendix C contains the evidence tables developed from the literature synthesis.

Chapter II. Methods

II. Methods

This chapter documents procedures that the RTI-UNC Evidence-based Practice Center (EPC) used to develop this systematic evidence review (SER) on cervical cancer screening. We document the general search strategy and criteria, specify the final search terms, describe procedures for further review of abstracts and publications, explain methods used to extract information from included articles, introduce approach to summarizing findings, and detail the peer review process.

During preparation of the evidence report, EPC staff collaborated with two current members of the US Preventive Services Task Force (USPSTF). Our collaboration took place by conference call, electronic mail, traditional mail, and fax. We presented steps of the development of this SER at USPSTF meetings in February, May, and September 2000. A key consideration for the September 2000 meeting was the timing of the report with respect to the anticipated publication in the near future of results from the first randomized trial comparing cervical cancer screening tools.

Literature Search Strategy

Our key questions guided the preliminary review of the literature. We have emphasized the identification of new research, existing syntheses of the literature, and opinions of leading medical and policy organizations, especially those reported since the completion of the second *Guide to Clinical Preventive Services*.[4] As part of the preliminary search, we took four steps: (1) reviewed prior USPSTF findings; (2) obtained current recommendations and/or guidelines for cervical cancer screening from the American Academy of Family Physicians, American

Chapter II. Methods

Cancer Society, American College of Obstetricians and Gynecologists, Australian Health Ministers' Advisory Council, Canadian National Workshop on Screening for Cancer of the Cervix, the National Strategic Plan for Early Detection and Control of Breast and Cervical Cancer, and the UK National Health Service Cervical Cancer Screening Program; (3) identified recent relevant systematic reviews in the medical literature; and (4) consulted with USPSTF liaisons for this topic.

Inclusion/Exclusion Criteria

We established overall inclusion and exclusion criteria a priori. Tables 3 and 4 present results of the search strategy with specific search terms.

Our search strategy, developed with the assistance of the RTI-UNC EPC research librarian who specializes in evidence-based literature review is described in Table 4. Using a selection of sentinel publications relevant to each key topic that were captured in the original broad search (Table 3), we specified searches that would provide focused identification of articles related to each key question. However, further specifying exhaustive searches for each question resulted in oversight of articles likely to be relevant as judged by missing sentinel articles. Key Question 1 about older age, older age and interval, and hysterectomy was the most difficult search to focus. As a result, we took an exhaustive approach to categorizing all articles obtained in the larger search. This process is described in detail below.

Chapter II. Methods

Literature Reviewed

Citation Database

We used the search strategy in Table 4 to identify potentially relevant publications. From our first search (December 1999) we imported references into a ProCite database, which enumerates, stores, manages, and retrieves bibliographic citations; we also recorded the fate of all identified publications as they were screened for inclusion. We repeated the search in June 2000, eliminating duplicate references and adding the new additions to the ProCite database.

Screening of Articles

Two EPC staff independently reviewed the titles and abstracts of the articles identified and excluded those that did not meet eligibility criteria. If the reviewers disagreed, we carried the article in question forward to the next stage in which we then reviewed the full article and made a final decision about inclusion or exclusion. At each step we recorded the fate of the article in the ProCite database. Table 5 presents the disposition of articles identified as potentially relevant publications (for review of the full article), summarizing the number of publications at each step and their categorization.

Limiting the exhaustive search (search 8, Table 4) to identify only articles that reported on trials, we identified 57 articles. Of these, 25 are primary reports of randomized trials: 15 address methods to promote uptake and continuance of appropriate screening; 3 examine methods to improve follow-up of abnormal screening findings; 3 compare tools for collecting cytologic samples (i.e., type of spatula, brush or swab); 3 investigate patient education and

Chapter II. Methods

satisfaction; and 1 compares cytology alone to cytology with cervicography as a primary screening modality. This exercise confirmed our assessment that few data would be available from randomized controlled clinical trials to inform our review.

Because final inclusion criteria were closely linked to the intent of the key question, we give greater detail about selection of articles for each of the key questions below. Table 5 summarizes the disposition of the articles identified in the literature search and the number of full articles on each topic retained for review.

Key Question 1: Screening Among Older Women and After Hysterectomy

The majority of the relevant literature identified for Key Question 1 is based on site-of-care or population-based prospective and retrospective studies and case-control studies. The large number of full articles that we retained for final abstraction reflected the need to examine the articles themselves to determine whether they provided data on benefits and risks of screening for particular groups (i.e., women age 65 and older) and in sufficient detail to clarify the relative performance of different screening intervals. We specifically sought articles that evaluated clinical risk prediction tools for assigning screening interval. Samples of the screening forms used to document decision making about inclusion of articles for Key Questions 1A and 1B appear in Appendix B.

In total, we screened 118 full articles to determine relevance. Of these, we retained 42 articles for Key Question 1A and 1B: 14 for full abstraction and 28 for supplementary information.

Chapter II. Methods

Key Question 2:   New Technologies for Cytology

Although our search strategy identified articles since 1995 relevant to Key Question 2 on new methods for preparing or evaluating cytology, we conducted full abstraction of only those publications that had appeared since completion of the *AHCPR Evidence Report/Technology Assessment, Number 5: Evaluation of Cervical Cytology* prepared by a team at the Duke Medical Center EPC.[32] Working with colleagues at the Duke EPC, we applied criteria and data extraction techniques that result in comparable reporting formats between the two evidence reviews. This approach allows us to streamline preparation of this report while capitalizing on the data previously collected to describe the performance characteristics of new cytology technologies.

The *Evaluation of Cervical Cytology* evidence report had several goals, including developing a model to estimate the costs associated with cervical cancer screening, evaluation, treatment, and follow-up of cervical cytologic abnormalities and the costs of treatment and follow-up of cervical cancer. To accomplish their modeling goals and make the best use of the existing literature, the Duke EPC included studies in which relative performance of new technologies could be assessed (e.g., use of computerized rescreening compared to conventional cytology alone) through use of expert adjudication of cytology findings with clinical confirmation of at least 50% of high-grade lesions.

Our team elected in advance of the literature review to focus primarily on studies that used clinical confirmation of cytology by colposcopy, cervical biopsy, or both (see Key Question 2 Screener in Appendix B). Specifically, we required that the new method being evaluated be

- Obtained as a screening test or adjunct to screening (i.e., not as follow-up of documented disease);

Chapter II. Methods

- Compared with a reference standard of histology or colposcopy;
- Verified by use of the reference standard within an average of three months interval from the screening sample; and
- Reported in a fashion that allows completion of a 2x2 table relating the new method to the reference standard.

In the case of new Pap testing technologies, these strict criteria also required that performance of the new test could be compared to that of conventional Pap testing.

We screened 48 full articles to determine relevance for updating Key Question 2. Overall, based on these strict inclusion criteria, only three articles were eligible for inclusion. Virtually all excluded articles failed to use a clinical reference standard or did so only among those with a positive test without confirmation of either all or a subset of test negative tests. Because a secondary goal of our work was to update the *Cervical Cytology* report, we relaxed our criteria to include studies that would have met the criteria for that report (which allows for a cytology reference standard). In total, we abstracted three high-quality articles and five other articles for the evidence tables and retained five for supplementary information.

For completeness, we added these new studies to the new technologies evidence table originally produced for the *Evaluation of Cervical Cytology*. The updated table is reproduced in its entirety in this report as Evidence Table 2, Appendix D. We did not identify any studies that report rates of outcomes over time, such as rates of cervical dysplasia or cervical cancer among cohorts of women receiving screening using conventional cytology compared to either new technologies alone, or to a system that adds a new technology to conventional cytology. We found no publications of randomized trials comparing individual outcomes by screening modality.

Chapter II. Methods

Key Question 3: What is the Role of HPV Testing?

After consultation within our team and with the USPSTF in a February 2000 conference call, we restricted this portion of the review to recent literature. Although the previous *Guide to Clinical Preventive Services* recommended against use of HPV testing as an ancillary test within a screening strategy, we believed that the laboratory technology for HPV identification, classification, and measurement of viral DNA incorporation has changed so rapidly that a review of the recent literature was appropriate.

Our search identified 64 abstracts potentially relevant to this topic; once again the individual review of all abstracts identified more articles for full review than restriction of search #8 in Table 4 using a variety of electronic search strategies. In total, we screened 30 full articles to determine relevance. The strict inclusion criteria for assessing screening studies were applied to the HPV literature. Studies on this topic were more likely to include histologic verification of test results, although no prospective comparisons of individual patient outcomes or published randomized trials examine the efficacy or effectiveness of HPV testing in clinical use. Overall, 16 of these 30 articles were included for this question, 13 for full abstraction and 3 for supplementary information.

Cost and Harms

The screening worksheet for each key question included additional questions to identify articles that reported on cost, cost-effectiveness, and harms, including psychological distress. When we flagged an article as potentially providing background on cost, cost-effectiveness, or other relevant economic evaluations, we subjected the article to further review by another team member.

Chapter II. Methods

We reviewed 21 articles, the vast majority of which were articles that examined the cost of operating national screening systems within single payer health systems across Europe. As our goal was not cost evaluation or modeling, or critique of that literature, we wished to retain only the most relevant publications for this SER. To be considered, we required that the article reflect the cost or charges of using the screening modality in a primary care setting in the United States and that it compare new methods with conventional methods. We included four publications relevant to cost, including an evidence report from the Agency for Healthcare Research and Quality (AHRQ), as background references.

In consultation with our team members, USPSTF liaisons, and USPSTF members in telephone conference, we elected not to examine relative rates of assignment of the diagnosis of atypical squamous cells of uncertain significance (ASCUS) and cervical intraepithelial neoplasia I (CIN1) as a potential indicator of "harm." Many such low-grade lesions regress and watchful waiting may be as effective as intervention; however, evidence at this time is insufficient to assert that detection of low-grade changes confers no benefit and only poses potential harm. Nonetheless, we have emphasized that optimal screening systems maximize identification of high-grade changes. Similarly, we have noted the potential harm of undergoing unnecessary evaluation and/or procedures, whether for diagnosis (such as colposcopy) or treatment (such as loop electrosurgical excision procedure or LEEP), but we have not further examined the harms of these procedures themselves.

Chapter II. Methods

Literature Synthesis and Preparation of Systematic Evidence Review

Data Abstraction and Development of Evidence Tables

We abstracted information about study objective, design, population, conduct, outcomes, and quality into designated sections and positions within evidence tables created in Microsoft Excel and Word. Two readers, a methodologist and a clinician-researcher, reviewed each article in an evidence table. The order of review by each pair of readers was not mandated, and both parties checked calculations of summary data, such as test sensitivity, that was generated for the tables.

To assess systematically comparable features of included articles and assure consistency, we used a checklist of potential indicators of study quality for the literature related to each key question. For Key Questions 2 and 3, we provide scores using the system designed for the *Evaluation of Cervical Cytology* evidence report, which fully documents development of the scoring system.[32] For Key Question 1, we incorporated indicators more relevant to cohort research, eliminating those items related purely to evaluation of diagnostic tests. Scores were assigned separately by two individuals and discussed as a group in the rare cases of substantial differences of opinion. These scores and a global categorization of the internal and external validity of the reviewed research contributed to grading of individual articles and the body of relevant literature consistent with USPSTF methods.[33]

Chapter II. Methods

Peer Review Process

Upon completion of the draft SER incorporating the review at the September 2000 USPSTF meeting, we conducted a broad-based external review of the draft. Among the outside reviewers are representatives of key primary care professional organizations that have formal ties to the USPSTF, a representative of the Canadian Task Force on Preventive Health Care, representatives of other professional societies, clinical experts in the area of cervical cancer screening, members of the AHRQ staff, and representatives of other federal agencies. Appendix A lists the names and affiliations of all peer reviewers. We took into account the comments of these reviewers in developing the final version of this SER.

Chapter III. Results

III. Results

This chapter presents the results of our systematic review on three main issues: screening among women who are 65 years of age and older or who have had a hysterectomy, technologies for cervical cytology, and testing the human papilloma virus (HPV) as a part of cervical cancer screening. Evidence Tables detailing information from the literature examined by members of the RTI-University of North Carolina Evidence-based Practice Center (RTI-UNC EPC) can be found in Appendix C.

Key Question 1

Screening Among Women Age 65 and Older

Twelve articles since 1995 provided sufficiently detailed information about results of screening by age to examine the evidence about screening among older women (Evidence Table 1A, Appendix C). For inclusion, we required that the study include women above age 50, that data be presented stratified by age or in subanalyses that compared older to younger women, and that the denominators for the outcomes be known (e.g., the number of abnormal Pap results reported corresponds to a defined number of individual women screened or number of Pap tests obtained in a specified group of women). The majority of rejected studies were ecologic-level reports correlating population-based rates of detection of dysplasia and/or cancer with Pap testing trends, or case-control studies that matched on age and contributed only to the literature about interval, or models validated using data that would not meet the inclusion criteria.

Chapter III. Results

In toto, the included studies tell different versions of the same story about the influence of age on risk for high-grade cervical lesions and cancer:

- Incidence of cervical intraepithelial neoplasia (CIN), including CIN 3 and carcinoma in situ (CIS), peaks in the mid-reproductive years and begins to decline in approximately the fourth decade of life.[26,34-36]

- The prevalence of CIN follows a similar pattern: diagnosis of CIN 3 and CIS is shifted toward older age at diagnosis relative to CIN 1 and 2 but still decreases with age.[37-40]

- This general pattern is also apparent among previously unscreened women.[40,41]

- Based on the incidence of cancers that arise between screening intervals, cervical cancer in older women is not more aggressive or rapidly progressive than that in younger women.[42]

- The rates of high-grade squamous intraepithelial lesions (HSIL) diagnosed by cytology are low among older women who have been screened: 0.2-1.9,[36] 0.7-1.7,[40] and 0.8-1.4,[26] per 1,000 among women ages 50 years and older.

These observations are consistent with ecologic data and natural history studies of cervical dysplasia and cancer and studies of HPV progression. Three of the included studies rely only on cytology results.[26,36,37] The remainder have varying degrees of histologic documentation; these range from complete or near complete histologic documentation of all cases,[38,40-44] to good verification of cases (72% verified[39] to 82% verified[35]) to inadequately described use of histologic "gold standard."[34] None attempted to identify false negatives through evaluation of women with negative screening tests. As such, this literature reflects the quality of the cytology services deployed to screen the retrospective and prospective cohorts that make up

Chapter III. Results

the study populations. Relative comparisons of the effectiveness of the screening systems and of the potential impact of screening interval cannot be made across studies.

With these caveats, we describe in greater detail selected studies of incidence of histologic or cytologic abnormalities among previously screened women, prevalence within screening programs, and the influence of interval and prior screening history on probability of detecting abnormalities.

Incidence and Age

Each of the studies reporting incidence of CIN and more serious lesions constructed their cohort by requiring a normal Pap test as the earliest reference point. Computerized record-keeping systems from a single cytology laboratory[34,35] assure relative consistency of quality of examination of the baseline and subsequent Pap tests; in relatively stable populations, they also allow exclusion of women with prior documentation of abnormal cytology. Sawaya and colleagues prospectively gathered data assuring that baseline cytology for individuals entering the study was reported as normal; however, this does not exclude the possibility that participants may have had a more remote history of dysplasia evaluation or treatment.[26,36,44] In the reports based on the National Breast and Cervical Cancer Early Detection Program, laboratory facilities across the United States processed Pap tests;[26,36] in the Heart and Estrogen/progestin Replacement Study (HERS) study, a single commercial laboratory (Empire Pathology Medical Group, Garden Grove, CA) processed tests.[44]

Despite these distinctions and different groupings of age in the analysis of data, estimates of incidence of high-grade lesions are compatible across the studies of incidence: 1.0 per 1,000 women 41 years of age and older, screened at an interval of approximately 1 year;[34] 1.2 per

Chapter III. Results

1,000 person-years among women 40 years of age and older;[35] 1.4 per 1,000 women ages 50 to 64 years and 0.8 per 1,000 women 65 and older screened at an interval of 36 months or less;[26] 0.6 to 1.9 per 1,000 smears in women ages 50 to 59 years; 0.2 to 1.2 in women 60 or older;[36] and 0.4 per 1,000 women in a post-menopausal cohort with average age of 66.[44] Fewer than one woman in a thousand (in some studies as few as two to six in 10,000) who were age 60 and older and had a negative smear at baseline received a new diagnosis of CIN 3 or cancer. In studies that had cases of cancer in screened populations of older women, the ratio of CIN 3 to cancer range from 4 to 1[26] to more than 10 to 1.[40]

Prevalence

Studies of prevalence (i.e., those that include screened and unscreened women without requirement of baseline normal cytology) are also compatible with the assertion that risk of high-grade cervical abnormalities decreases with age. The only study that does not appear to support this view is that of Formso and colleagues;[37] they report 2.23 cytologic findings of CIN 3 per 1,000 women age 50 to 59 years; 1.96 per 1,000 for women age 60 to 69 years; and 4.24 for women ages 70 and older. The inclusion criteria specified "no prior abnormal Pap test"; this allowed inclusion of women having their first screening ever or after an extended interval. In the text, the authors noted the large proportion of older women, especially those older than 70, who had first Pap tests included in these data. Their data are based on records from the same cytology laboratory in Norway used by Gram et al.[35] The laboratory serves a relatively stable population in two northern counties without a formal screening program. The time periods in the analyses by Formso et al.[37] and Gram et al.[35] overlap by two years; thus, presumably individual-level data may be duplicated in these reports. Gram and colleagues sought histologic

Chapter III. Results

confirmation of CIN 3 and carcinoma and achieved an 82% verification rate, whereas Formso and colleagues relied on cytology alone. The differences between their findings are certain to reflect the inclusion of unscreened women in the prevalence study[37] but not in the incidence study.[35] The differences may also reflect discrepancies between cytology and histology among older women and unstable estimates based on small numbers.

The remaining publications reflect falling risk with age and low prevalence among older women. These include the Cecchini et al. study of previously unscreened women in Italy,[41] Cruickshank's study[39] among British women who are actively advised to be screened every 3 years, and the work by Lawson and colleagues based on opportunistic screening in the US National Breast and Cervical Cancer Screening Program.[40] Among American women being screened for the first time in the study by Lawson et al.,[40] rates of CIN 3 or cancer were 2.3 per 1,000 Pap tests among women ages 50 to 64 and 1.7 per 1,000 among those ages 65 and over, followed by a reduction to 1.3 per 1,000 and 0.7 per 1,000 at the second screening opportunity.

Screening History and Interval

The significance of prior negative smears is directly addressed by both age-adjusted and age-specific analyses in the examination by Sawaya and colleagues of the probability of abnormal Pap after 1, 2, or 3 prior normal Paps. The probability of HSIL decreases with each subsequent normal smear in women ages 30 and older. Reduction within age brackets is detailed in Evidence Table 1. Investigation of the influence of interval suggests that longer intervals are associated with similar or increased detection of high-grade lesions;[26,35,43] other work not reviewed here (but considered by the US Preventive Services Task Force [USPSTF] in the mid-

Chapter III. Results

1990s) suggests that cancer outcomes are statistically comparable with intervals of three to five years. Three years was recommended as a conservative estimate of appropriate interval.

Screening among Women Who Have Had a Hysterectomy

The 1996 USPSTF report is unequivocal: "women who have undergone a hysterectomy in which the cervix was removed do not benefit from Pap testing, unless it [the hysterectomy] was performed because of cervical cancer."(p.111)[4] This opinion was based on recognition that Pap testing in the absence of a cervix no longer constitutes screening for cervical cancer. In this context the Pap test becomes screening for vaginal cancer, a yet more rare condition. Prior publications support this view.[45] Nonetheless, a recent publication based on practice patterns at the Marshfield Clinic (a 450-physician multi-specialty clinic in Wisconsin) suggests that more than half of women who have had total hysterectomies for benign disease continue to receive screening at an average of one test every 3.5 years.[46] This confirms the clinical opinion of members of our group that the majority of women continue to receive screening even after hysterectomy for benign disease.

Two additional studies documenting the low risk of cytologic abnormality after hysterectomy have been published since the 1996 USPSTF recommendations. The first, among women age 50 and older, is a cross-sectional study with a nested case-control component.[47] They documented that identification of dysplasia and cancer was rare (1.6/1,000 tests) in this age group, especially after hysterectomy (0.18/1,000). They also showed that, compared to matched controls, women after hysterectomy were one tenth as likely as those with a cervix to have any Pap test diagnosis of abnormality.[47] Likewise, Pearce and colleagues' study of 6,265 women (with 9,610 Pap tests) who had hysterectomies for benign disease found a total of 104 abnormal

Chapter III. Results

Pap test from 79 women within a two-year timeframe.[48] At completion of follow-up, 3 women had vaginal intraepithelial neoplasia (VAIN) and one had squamous cell carcinoma. This translates into 0.42 high-grade lesions per 1,000 Pap tests.[48] These articles are described in greater detail in Evidence Table 1B.

Summary

The available evidence about age and screening outcomes is observational, from large population-based retrospective cohorts supported by registry systems, or from smaller prospective cohort studies. Despite varied study design and populations, the findings are coherent and support the assessment that risk of high-grade cervical lesions falls with age, especially among those with prior normal screening results. Because none of these studies evaluates outcomes among women who did not receive further screening after a designated age, and because none is experimental, we cannot draw direct conclusions about the anticipated results of discontinuing screening at a specific age.

Prior USPSTF recommendations to discontinue Pap testing after hysterectomy for benign disease are supported by a well-conducted study; they should be re-emphasized. Lastly, no direct comparisons between proposed screening systems, including those based on individual risk assessment, were identified. This confirms that this literature is substantially less well-developed than that of other areas of screening tests, such as colorectal cancer screening in which trials comparing annual fecal occult blood testing to sigmoidoscopy are available to inform decisionmaking.

Chapter III. Results

Key Question 2

New Methods for Preparing or Evaluating Cervical Cytology

Key Question 2 (To what extent do new methods for preparing or evaluating cervical cytology improve diagnostic yield compared to conventional methods?) is addressed predominantly by direct comparisons of diagnostic tests. The majority of the literature identified for Key Question 2 is based on archived laboratory specimens. These studies compare the techniques being evaluated with the results of review by a panel of cytology experts. Most often, these comparisons are conducted by subjecting specimens with a pre-selected mix of normal and abnormal cytologic results, to review by the techniques under study. In general, discrepancies between cytology reports were adjudicated by the expert panel masked to findings; rarely, subsets of normal or concordant diagnoses were also reviewed by the panel.

Very few studies of new technologies are validated by concurrent or subsequent colposcopy or histology of abnormal screening test results; even fewer include validation of normal screening test results. This means that in almost all studies identified the sensitivity, specificity, and predictive values of the technology cannot be directly assessed or compared with the test characteristics of conventional cytology in the same population.

Our search identified 196 articles on the types of technology we wished to review. We excluded 143 of these articles at the time of abstract review because they did not meet basic inclusion criteria (e.g., were commentaries, were based on experimental laboratory systems, did not have human subjects). Forty-eight full articles were retrieved; these included 23 that were found not to meet basic inclusion criteria (commentary and reviews). Of the remaining 25 articles, four were relevant new articles not previously abstracted and summarized in the *AHCPR*

Chapter III. Results

Evidence Report/Technology Assessment, Number 5: Evaluation of Cervical Cytology; two were new articles on screening tools not covered by either systematic review (e.g., unaided visual inspection of the cervix, and cervicography); and one article was a final publication of a study that had been reviewed in draft form for *Cervical Cytology* report and had minor changes in the final published results.[49]

The *Cervical Cytology* report was published in February 1999[32] and updated in January 2000 for peer-reviewed publication.[50] The report includes studies with and without validation of screening tests by clinical evaluation. For convenience, the portions of the evidence tables from that report that are related to new technologies are reproduced and updated in this report (Evidence Table 2). In collaboration with the authors of the cervical cytology report, the RTI-UNC EPC team abstracted new articles in a comparable fashion.

As described in the methods section, we had planned to focus on studies that used colposcopy or histology as a gold standard for evaluation of performance of the screening system. However, such publications remain rare. At completion, we extended the prior *Cervical Cytology* review by updating final data from one article, adding three new articles based on a cytological reference standard with or without a subset of histologic verification, and adding one with an adequate colposcopic and histologic reference standard. This study applied a definitive clinical reference standard to a random sample of women with normal screening test results and permitted calculation of all test characteristics including estimation of specificity.[51]

Evidence Table 2 (Appendix C) summarizes 29 studies that evaluate the performance of new technologies for preparing or interpreting cervical cytology specimens: 9 evaluating liquid-based cytology collection systems (ThinPrep®); 13, neural-net rescreening or prescreening

Chapter III. Results

studies (PapNet®), and 7, computer algorithms for selecting slides for rescreening or for screening (AutoPap®).

A single study evaluating ThinPrep® liquid-based cytology and other screening modalities in a population-based cohort of 8,636 Costa Rican women met strict review criteria for prospective evaluation of the test characteristics of a new screening technology:[51]

1. Cytology specimen obtained as a screening study;
2. Performance of new technology was evaluated by colposcopy and/or histology as the reference standard;
3. Pap and reference standard were applied within an average of three months; and
4. A validation of normal cytology results was undertaken, such that a complete 2x2 table relating the new technology to the reference standard can be completed.

The study that met these criteria used colposcopy and histology to evaluate women in three categories: (1) atypical squamous cells of uncertain significance (ASCUS) on cytology as assessed by ThinPrep, PapNet, or conventional smear; (2) positive cervigram; or (3) physical examination suspicious for cancer or gynecologic emergency (rare indication). These investigators conducted a validation substudy among a random sample of 150 women with negative screening results, thus allowing estimation of specificity and predictive values. Their reference standard was a composite of all screening tests and histology findings, called the final case diagnosis. The final case diagnosis categories were negative (including ASCUS with normal colposcopy and/or histology), low-grade squamous intraepithelial lesion (LSIL) (by histology, or by cytologic confirmation by two or more methods), HSIL (93% histologic confirmation), invasive carcinoma (100% histologic confirmation), and equivocal. Equivocal included women with a single cytologic diagnosis of LSIL, isolated positive cervigram, or

Chapter III. Results

conflicting results on the basis of all the data. Because the threshold for colposcopy was ASCUS or higher, the majority of equivocal final case diagnoses are presumed to be equivocal low-grade findings although the precise make-up of this group is not clearly specified in the article.

Bearing this classification system in mind, we have calculated performance measures for ThinPrep® using two different assumptions about equivocal final diagnosis; one assumption includes them in normal, and the other in LSIL. Based on the findings of no abnormal histology among the random sample of 150 women with normal final diagnoses (i.e., all screening test negative) who were evaluated with colposcopy and biopsy, estimates of sensitivity, specificity and predictive value are based on the presumption of no false negatives, which is plausible based on the number of tests applied and the consensus process for assigning the final diagnoses. Table 6 summarizes test characteristics at different thresholds; assignment of the equivocal cases is indicated.

Two studies of ThinPrep®, known for the large size of the populations screened (>35,000 each), did not meet strict criteria. These studies provide histology results for a subset of subjects obtained within undisclosed periods of time from the screening ThinPrep®.[52,53] Neither used a colposcopy/histology reference standard to verify test negatives, and both studies appear to rely on histology specimens associated with clinical care for calculating sensitivity and specificity of the ThinPrep® test compared to conventional cytology.

Among 10,694 US patients in the latter study screened by ThinPrep®, 630 women had cytology diagnoses of ASCUS or more severe; 1,395 of 39,408 women screened by conventional cytology had cytology of ASCUS or more severe.[53] However, the estimates of sensitivity and specificity reported in the publication are based on 54 biopsy reports in the ThinPrep® group and 89 in the conventional cytology group. The authors do not specify how these "available"

Chapter III. Results

biopsies were identified. They could not capture those women who had normal colposcopy, and therefore no biopsy, among those with abnormal cytology. Additionally, ThinPrep® and conventional specimens were from separate groups of women, not one of each specimen type from each individual. As a result, the sensitivity and specificity of ThinPrep® (reported as 95% and 58%, respectively) are not valid and cannot be appropriately compared with conventional cytology (reported as 85% and 36%).

The earlier study obtained split samples for ThinPrep® and conventional cytology from a cohort of 35,560 Australian women.[52] Within the Australian health care system, colposcopy is the recommended immediate evaluation for "inconclusive" slides in which high-grade abnormalities cannot be excluded and for cytology diagnoses of CIN1 and above. High-grade or inconclusive cytology results were reported for 433 ThinPrep® specimens and 430 conventional cytology specimens, of these, 325 (75%) and 319 (74%), respectively, had histology results. This again excludes women who had colposcopy without biopsy. However, the focus on high-grade lesions makes the probability of biopsy high, and so the proportion with follow-up is adequate. In this context, the relative true-positive and false-positive rates of the test can be used to compare performance.[54] The relative true-positive rate for ThinPrep® compared to conventional cytology for detecting high-grade histologic abnormalities is 1.13, suggesting a modestly higher sensitivity of Thin Prep®; the relative false-positive rate is 1.12, suggesting a modestly lower specificity.

Neural-network Rescreening

The literature about neural-network rescreening (and screening) technology is also fundamentally limited by the lack of histologically confirmed performance measures. Applying

Chapter III. Results

the same criteria outlined above, we found no studies that met the inclusion standards. One study of 160 slides originally classified by conventional cytology as ASCUS and subsequently scanned by the PapNet® system is supported by histology for all samples.[55] This study failed to meet criteria because the reference standard was obtained up to one year after the screening test and because the slides reviewed do not reflect a representative distribution of normal and abnormal specimens. We summarize this work and two other studies with some use of histology documentation as examples of the content of the literature.

The slides for this ASCUS review study, selected because the patients had subsequent histology on record, were re-read using PapNet®. Although the spectrum of disease is skewed and reflects a high prevalence of abnormalities, performance measures can be calculated for LSIL and HSIL given an initial ASCUS determination by conventional Pap. If histology of HPV changes only is separated from CIN 1 and included with normals, the sensitivity, specificity, positive predictive value, and negative predictive value of PapNet® categorization of LSIL or greater are 45.4%, 85.8%, 45.5%, and 85.8%, respectively. If HPV changes are included in CIN, these values are, respectively, 37.7%, 92.3%, 78.8%, and 66.1%. At a PapNet® threshold of HSIL or greater for detecting CIN 2, CIN 3, or carcinoma, the sensitivity was 27.3%, specificity was 94.2%, positive predictive value, 42.9%, and negative predictive value, 89.0%.

Jenny and colleagues hand-selected a slide set consisting of 516 abnormal slides with accompanying histologic record of abnormality and mixed them with 684 slides from women with 2 years of normal follow-up.[56] The study was designed to evaluate the consistency of reports between two independent PapNet® evaluations and a manual screening of the same slide set. The PapNet® evaluations identified 91% and 84% of the histologically proven abnormal specimens; conventional screening classified 78% as abnormal. The resulting true positive ratio

Chapter III. Results

of PapNet® to conventional screening ranges from 1.08 to 1.16, supporting the claims that PapNet® reduces false negatives. Unfortunately, the research team missed the opportunity to evaluate all performance measures using the 2-year normal follow-up group as a fairly well-documented confirmation of test negatives. The authors do not comment on the classification of the normal slides, focusing only on the 516 slides from women with known abnormalities.

Using data from the centralized Dutch national pathology reporting program, Kok and colleagues evaluated cervical cancer outcomes and risk of screening method failure in 2 large groups: 109,104 smears evaluated by conventional screening and 245,527 slides submitted for neural-network based screening (not rescreening as in the United States) with PapNet®.[57] They used the reporting system to identify 71 women with a diagnosis of biopsy-confirmed squamous cell carcinoma and to locate the screening smear obtained prior to the biopsy. Of the 71 smears, 19 had been evaluated by conventional methods and 52 by PapNet® screening; there were four false negatives among the conventional screening group and five false negatives among the PapNet® slides. The balance of the 71 smears were all reported with a level of abnormality ranging from LSIL to carcinoma that prompted clinical follow-up. The false negatives were subjected to masked detailed review by both standard microscopy and PapNet®. All five PapNet® failures were confirmed to be sampling errors; i.e., the slides did not contain identifiable abnormal cells as assessed by hand review or PapNet®. The conventional screening failures did contain scant abnormal cells and constitute screening failures. Two independent, masked PapNet® re-evaluations of the slides classified three of the four as "suspicious" and the fourth as CIN 3, suggesting that the screening failures could have been averted because none of these individuals would otherwise have been returned to the 5-year routine screening interval.

Chapter III. Results

Computerized Rescreening

AutoPap® is a computerized quality control system that can be used for initial screening and for optimizing selection of high-risk slides for rescreening. Its current routine use in the United States is restricted to rescreening. We found no studies that met strict criteria for review, although we did identify two studies by the same group that employed a level of histologic verification of test performance. The first was based on an 86-slide set of smears from histology-proven HSIL cases.[58] This study retrospectively measured whether the system "found" these slides when they were reviewed as part of a larger study. The system can be set to determine the proportion of slides that will be rescreened; this study evaluated 10-percent and 20%- rescreening thresholds. At the 10%- review threshold, AutoPap selected 77% of the HSIL slides for assessment; at the 20%- threshold, it selected 86% for assessment. These figures are substantially higher than the expected maximum selection rate of 10% of false negatives if the standard random sample of 10% of negative slides were to be used to determine rescreening.

In a subsequent study, this group focused on ability of the AutoPap® system to aid identification of HSIL lesions.[59] This cytologic diagnosis should consistently be associated with clinical evaluation and histology, which improved their ability to confirm positive diagnoses although again they have no verification of the test-negative specimens. Within a parent study with 25,124 screened slides, they identified 70 slides supported by definitive diagnoses of HSIL, CIS, or invasive cancer. They compared the detection rate for these slides in the convention practice screening and the AutoPap® screening portion of the parent study. Both modalities detected 63 of 70 abnormal slides; conventional practice failed to detect five and AutoPap® failed to detect two. The performance of the two modalities was statistically equivalent.

Chapter III. Results

Although we updated the evidence tables from the *Cervical Cytology* report by including two new studies about PapNet[57,60] and a final publication of an article available in manuscript form,[49] we have not summarized them in greater detail here. The reason is that these reports use cytology reference standards and do not change the overall picture of the performance of new technologies. Thus, they serve as confirmation of similar earlier publications, not as wholly new information.

Harms

We did not identify publications that specifically address harms of new technologies for cervical cancer screening. These screening tools are implemented at the laboratory level and not at the level of clinical specimen collection, so they do not increase risk of harm from the actual specimen acquisition.

The performance characteristics of the new technologies will determine the risk of harm. Although the data are limited, on average these tools improve sensitivity and reduce specificity. This finding suggests that increased detection of low-grade lesions and false positives are the primary potential sources of harm; i.e., harm may take the form of increased evaluations, possible over-intervention, and psychological distress for the women diagnosed with abnormalities. These harms are poorly documented for conventional Pap testing and have not yet been assessed for new technologies.

Benefits

Likewise, direct benefits of new technologies for improving care processes or outcomes are not documented. The characteristics of these new tools suggest that they can improve

Chapter III. Results

detection of precancerous lesions when compared in a single screening session to conventionally prepared and interpreted cytology. However, no data are available to assess their long-term benefit if implemented in a screening system. Cervical dysplasia is at times a spontaneously resolving condition or, if progressive, a slowly developing condition. For precisely these reasons, the full system of screening—i.e., the modality used, the interval for testing, and the decisionmaking process related to evaluation and treatment—need to be evaluated in toto to compare performance characteristics properly. Prospective measurement of outcomes is essential to guiding policy.

Costs

A central goal of the *Evaluation of Cervical Cytology* evidence report was modeling the effects on total health care costs, morbidity, and mortality of regular cervical cytologic screening using newer screening technologies compared with conventional Pap smear in women participating in screening.[32] Using a Markov model of a cohort of women ages 15 to 85, incorporating estimates about the natural history of HPV, and investigating one-, two-, and three-year screening intervals, they reached the following conclusions:

- The cost-effectiveness of either a technology that improves primary screening sensitivity (e.g., thin-layer cytology) or one that improves rescreening sensitivity (e.g., computerized rescreening) is directly related to the frequency of screening—longer intervals result in lower estimates of cost per life year saved.

- Findings were relatively insensitive to assumptions about cervical cancer incidence, cost of technologies, diagnostic strategies for abnormal screening results, age at onset of screening, or most of the other variables tested.

Chapter III. Results

- Substantial uncertainty surrounds the estimates of sensitivity and specificity of thin-layer cytology and computerized rescreening technologies compared with each other and with conventional Pap testing. This uncertainty is not reflected in the point estimates of cost-effectiveness. Although both thin-layer cytology and computerized rescreening technologies clearly improve effectiveness at higher cost, the imprecision in estimates of effectiveness makes drawing conclusions about the relative cost-effectiveness of thin-layer cytology and computerized rescreening technologies problematic.

- Given the uncertainty surrounding these estimates, all three technologies may well fall within accepted ranges of cost-effectiveness at 3-year screening intervals. No strategy or technology used for screening more often than every 3 years results in estimates of less than $50,000 per life-year.

This model substantially improves on prior work; it includes global costs of downstream care resulting from screening and cancers, more accurate estimates of the performance of conventional cytology then previously available, and sophisticated sensitivity analyses. However, important parameters of this model deserve note. Base assumptions include the following: (1) all women receive screening at the appropriate interval; (2) new technologies increase sensitivity without any decrement in specificity; (3) all patients receive appropriate follow-up; and (4) diagnostic evaluation of abnormal cytology detects all true abnormalities (i.e., no colposcopy or pathology errors are made). Adjusting each of these assumptions closer to actual clinical scenarios has the effect of increasing the cost-effectiveness ratio. If, as our update and the full *Cervical Cytology* report suggest, new technologies do have lower specificity than

Chapter III. Results

conventional cytology, then costs and harms of false positives have important system and individual implications.

The *Cervical Cytology* report also includes systematic review of prior literature on the cost-effectiveness of cervical cytology. In summary:

- Published models examining the cost and effectiveness of Pap smear screening have consistently found Pap screening to have a significant impact on the incidence and mortality of cervical cancer and to have an acceptable range of cost-effectiveness ratios when compared with no screening.

- Estimates of Pap test accuracy used in these models generally overestimated Pap test performance, as determined by recent unbiased studies, the findings of the report itself [*Cervical Cytology*], and a previously published meta-analysis. Best estimates of Pap test performance fall outside the range used in sensitivity analyses of some models.

Many of these models have results that are consistent when important parameters of the models are varied across of broad spectrum of assumptions. Ultimately, however, all current models are tied to the limitations in this literature and must be considered temporary substitutes for prospective research.

Summary

Overall, the quality of this literature is limited for the purposes of making decisions about choice of screening modality in the US population. We identified no randomized trials or prospective cost-effectiveness comparisons in suitable populations. In the absence of studies that

Chapter III. Results

relate the findings at screening to outcomes, the linkages between comparative test performance are insufficient to judge the implications of preferentially using one system over another.

Key Question 3: Role of HPV Testing in Cervical Cancer Screening

We examined two potential roles for HPV testing in cervical cancer screening. The first is focused on screening use of HPV testing, including simultaneous use with Pap testing. The second envisions a role for HPV in triaging women with abnormal screening Pap results with an emphasis on when knowledge of HPV status might modify decision-making about the need for or extent of further evaluation. Our specific questions were:

3A. Screening Use of HPV Testing: What are the benefits, harms, and costs of using HPV testing as a screening test, or of incorporating HPV testing at the time of the screening Pap test compared with not testing for HPV?

3.B. Triage Use of HPV Testing: What are the benefits, harms, and costs of using HPV testing as part of a screening strategy to determine which women with an abnormal Pap test should receive further evaluation?

We reviewed 30 potentially relevant publications. Ultimately, we abstracted 13 of these and entered their data into Evidence Tables 3A and 3B; we retained another three articles for background. The systematic review of the role of HPV testing in cervical cancer screening prepared by Cuzick and colleagues for the UK Health Technology Assessment Board of the National Health Service was a valuable aid.[21] Their publication influenced our decisionmaking about which testing modalities to consider appropriate for screening use, and it validated the

Chapter III. Results

literature search strategy we used for this SER by confirming that we were identifying appropriate publications.

Screening Use of HPV Testing

Six studies relevant to primary screening uses of HPV testing met four strict criteria for inclusion:

1. HPV specimen obtained and evaluated as a screening test;
2. Performance of HPV for predicting presence of CIN was evaluated by colposcopy and/or histology as the reference standard;
3. HPV test and reference standard were applied within an average of 3 months; and
4. A validation of negative HPV results was undertaken, such that a complete 2x2 table relating the HPV test findings to the reference standard can be completed.

These articles are summarized in detail in Evidence Table 3A. Table 7 extracts key features of all 6 studies and their populations to demonstrate the varied spectrum of disease in the populations screened. Although each of these studies used HPV testing in a population or clinic-based sample that enrolled consecutive women in a fashion comparable to screening use, 5 of the 6 studies used a study population at high risk for cervical dysplasia. One obtained specimens from HIV-positive German women;[61] 2 were based in Zimbabwe and enrolled primary care patients in a population chosen for high prevalence of HIV infection;[62,63] 1 selected participants in Cape Town, South Africa, with no prior screening;[64] and 1 enrolled women from Guanacaste Province, Costa Rica, where cervical cancer screening has been relatively unsuccessful in reducing cervical cancer incidence.[65] Only Cuzick and colleagues' study of 2,988 women

Chapter III. Results

having routine cervical cancer screening at 40 general practitioner practices approximates screening use in routine primary care practice in the United States.[21]

HPV types consistently considered high risk and tested by the assays used in these studies included types 16, 18, 31, 33, 35, 39, 45, 51, 52, 56, 58, 59, and 68. We have not considered performance characteristics of tests that detect only low-risk HPV types. Basic performance characteristics relating presence of high-risk HPV types to diagnoses are best summarized by considering HSIL and LSIL as separate diagnostic thresholds. For the HSIL threshold, we mean the ability of a positive HPV test to detect histology-proven HSIL or more severe lesions, such as invasive cancer. For the LSIL threshold, we mean the ability of a positive HPV test to detect histology-proven LSIL or more severe lesions, including HSIL.

HPV Screening for Detection of High-grade Cervical Changes

The sensitivity of a positive Hybrid Capture II test for high-risk HPV types for detecting histology proven HSIL or greater ranged from 62% to 95%.[21,62-65] Specificity ranged from 41% to 94%.[21,62-65] Details of test performance for detecting high-grade lesions, including predictive values and likelihood ratios for positive and negative tests, for each HPV test modality examined in these screening-based studies appear in Table 8.

Five of the 6 HPV screening studies used Hybrid Capture II.[21,61-64] To report aggregate estimates of performance for detection of HSIL, we calculated summary performance measures using the aggregate numbers of true positives, false negatives, true negatives, and false positives in these studies, with a combined total enrollment of 6,793 participants. In aggregate, the sensitivity of Hybrid Capture II for detecting HSIL was 84.4%; specificity was 78.7%; positive predictive value, 23.4%; and negative predictive value, 98.5%.

Chapter III. Results

These Hybrid Capture II estimates include an easily delineated group of women with HIV infection who had high prevalence of HSIL — 17.3% among the HIV-infected women in the Womack and colleagues cohort.[63] Excluding HIV-infected women from the aggregate test performance estimates should have the largest influence on the predictive values, because these are determined by prevalence. However, these women make up a small portion of the overall total and the estimates of performance characteristics are essentially unchanged. To further describe Hybrid Capture II performance in low-prevalence populations, we calculated aggregate data for only those studies[21,62,64] or arms of studies[63] in which the prevalence of HSIL was 10% or lower. In these conditions, sensitivity is somewhat lower at 81.6%; specificity is similar at 78.2%; positive predictive value is reduced to 18.2% and negative predictive value is 98.6%, essentially unchanged compared to the estimates that include all studies using Hybrid Capture II regardless of prevalence.

HPV Screening for Detection of Low-Grade Cervical Changes

Four of the screening studies include data that allow estimation of test performance if LSIL is included among the lesions targeted for detection by HPV testing. The sensitivity of a positive Hybrid Capture II test for high-risk HPV types for detecting histology-proven LSIL or greater ranged from 45.2% to 85.9%.[21,62,64] Specificities were between 47.0% and 96.7%.[21,62,64] Details of test performance for detecting low-grade and higher lesions are provided in Table 9.

Cuzick and colleagues study of women having screening at their primary care providers' offices in the United Kingdom is likely the most comparable to screening in the United States, the relative performance of the SHARP (Digene Corp.) microtiter format PCR test and the Hybrid Capture II is of interest. Both tests delivered similar performance: sensitivity of 64.3%

Chapter III. Results

versus 61.0%; specificity of 96.7% versus 95.1%; negative predictive value of 98.4% versus 98.6%, positive likelihood ratios of 19.4 versus 12.5, and negative likelihood ratios of 0.43 for both tests. In some studies, Hybrid Capture I delivers test performance comparable to SHARP PCR and Hybrid Capture II.[61,65] However, in the Cuzick et al. analysis (which allows direct comparison in the same cohort), the newer HPV testing modalities appear to be superior.

All of the studies that consider HPV and LSIL detection evaluated Hybrid Capture II. We calculated summary performance measures using the aggregate numbers of true positives, false negatives, true negatives, and false positives in these studies with a combined total enrollment of 5,674. In aggregate, the sensitivity of Hybrid Capture II for detecting LSIL or more severe lesion, was 67.3%; specificity, 80.1%, positive predictive value, 36.7%; and negative predictive value, 93.4%. Eliminating the HIV-positive women from the estimates,[63] the aggregate positive predictive value falls slightly to 35.7%; negative predictive value is essentially unchanged at 93.6%; sensitivity is 65.2%; and specificity is 81.2%. Lastly, eliminating all groups with a 10% or greater prevalence of LSIL or higher lesions, the resulting performance is sensitivity of 73.5%; specificity, 90.5%; positive predictive value, 29.2%; and negative predictive value, 98.5%.

Triage Use of HPV Testing

A proposed role for HPV testing is to serve as a diagnostic tool to facilitate triage of women with an abnormal Pap tests. To estimate its test characteristics in this role included 7 articles relevant to triage uses of HPV (Evidence Table 3B, Appendix C) that met the following criteria for inclusion:

Chapter III. Results

1. HPV specimen obtained or processed (or both) only for women with abnormal cytologic diagnoses on Pap testing;

2. Performance of HPV for predicting presence of CIN evaluated by colposcopy and/or histology as the reference standard;

3. HPV test and reference standard applied within an average of 3 months; and

4. A validation of negative HPV results, such that a complete 2x2 table relating the HPV test findings to the reference standard can be completed.

Table 10 summarizes the study populations and severity of disease identified in populations included in the studies relevant to HPV testing as a triage tool. Although the majority of these studies required an abnormal Pap test for enrollment, we required only that the HPV test and the colposcopy and/or histology, and not Pap testing, occur within an average of 3 months' time. Two of these studies were conducted in countries in which initial ASCUS and/or low grade cytologic findings are followed by repeat Pap testing. Colposcopy is only done if repeat cytology suggests risk of a high grade lesion.[66,67] This approach lengthens the time between the index Pap and the use of HPV testing and colposcopy. It also increases the probability that women undergoing colposcopy will have a histologic lesion. As a result of such cytology-based triage, the studies from Iceland and the Netherlands reflect a comparatively high prevalence of high grade lesions.[66,67] This finding has implications for this report because the prevalence of high grade lesions in these studies is higher than one would expect if HPV testing were used as initial triage at the time of a first abnormal Pap test, and prevalence does influence diagnostic test characteristics. These studies may not be as informative for US clinicians because within our screening approach, HPV is most attractive as a triage tool for guiding

Chapter III. Results

management of patients with low grade cytologic abnormalities who will at the time of colposcopy have a much lower prevalence of high grade lesions.

The indications for the colposcopy, including indications established for the purposes of research, vary across studies; this in turn also influences the spectrum of abnormalities identified. The study cohort in Iceland had a prevalence of 54.5% of high-grade lesions compared to 15.4% among German women referred for colposcopy. Similarly, among studies that enrolled patients based on cytology findings as an entry criterion (rather than referral for colposcopy), the criteria vary from ASCUS only[68] to CIN 1 or CIN 2.[66] This means that aggregating test performance characteristics across studies is problematic. The test characteristics of high-risk HPV testing for detecting HSIL as a triage tool are summarized in Table 11.

Three strong studies selected the degree of Pap test abnormality *a priori* and then obtained histologic verification of HPV test performance. The group with the lowest risk of high-grade lesion at the time of evaluation was that assembled by Manos and colleagues within the Kaiser Permanente Medical Care Program in Northern California.[68] They enumerated all women at 4 medical centers who had a Pap test during a 9-month period and invited all women with ASCUS to participate if they were not pregnant and did not have a history of treatment for cervical neoplasia in the prior six months. Sixty-one percent of patients with ASCUS enrolled in the study; participants were demographically similar to non-participants. In this setting the sensitivity of Hybrid Capture II testing for high-risk HPV to detect HSIL was 89.2%; specificity was 64.1%; positive predictive value, 15.1%; negative predictive value 98.8%. For detection of LSIL and more severe lesions, the sensitivity was 76.3%; specificity, 69.5%; positive predictive value, 37.8%; and negative predictive value, 92.4%.

Chapter III. Results

Herrington and colleagues studied 167 patients with 99 with persistent "borderline" smear, 39 with "wart virus changes", and 29 with "mild dyskaryosis."[69] Like women in the United States, German women with moderate or severe dysplasia are referred immediately for colposcopy and possible treatment. The group evaluated in this study is, therefore, much like the group for whom HPV triage would be of interest in the United States. In this setting, consensus PCR had sensitivity of 87.5%, specificity of 62.2%, positive predictive value of 42.2% and negative predictive value of 94.0% for detection of histology proven high-grade lesions.

Bollen and colleagues studied British women including only those with mild or moderate dysplasia or cytology.[66] Of the test systems they evaluated in their patient population, SHARP PCR (for high-risk HPV types) was relevant to this review. The sensitivity for detecting HSIL was 94.6%; specificity, 40.3%; positive predictive value, 39.8%; negative predictive value, 94.7%. In these 3 cytology-defined studies as well as the colposcopy-referral studies, the negative predictive value of testing for high-risk HPV types is greater than 90%, suggesting the feasibility of using negative HPV results to lengthen the time until repeat Pap test and to reduce the number of women who have colposcopy.

Benefits

The benefits of HPV testing as a screening test, an adjunct to screening, or a triage tool are not documented in prospective cohort studies or trials evaluating outcomes among women who received conventional cytology only compared to an alternate strategy that includes HPV testing.

Chapter III. Results

Risks

We did not identify any studies that explicitly addressed negative individual, health system, or societal consequences of HPV testing other than cost, which is addressed briefly below. Comments in various publications suggest potential risks fall within 6 categories of concerns listed below, but no literature addresses the likelihood of any of these reactions or their magnitude:

1. Focusing on HPV testing will strongly identify cervical dysplasia and cancer with sexually transmitted disease (STD) and will stigmatize those with the condition.

2. Identifying cervical dysplasia with a sexually transmitted virus will result in women who perceive themselves as at low risk for STD receiving less cervical cancer screening.

3. Diagnosis of HPV infection will provoke partner discord.

4. Diagnosis of HPV infection, given low-positive predictive value, will unnecessarily label some women as high risk.

5. Labeling women as infected with a high-risk HPV types will provoke anxiety that may have unpredictable consequences such as over-intervention or poor compliance with followup.

6. HPV testing could undermine the importance of cytologic screening with providers or patients.

Costs

Current costs and charges for HPV testing in clinical care vary widely by geographic location, test modality, laboratory contracts with vendors, and volume of specimens analyzed at

Chapter III. Results

the laboratory facility. At the University of North Carolina Healthcare System, the Hybrid Capture II test kit cost $43 in February 2001. The clinical laboratory fees were $125 for Hybrid Capture II testing and $160 for consensus PCR at the same time. As any other diagnostic modality, increased demand should eventually increase capacity, in part through competition, and lower costs. These market forces are not yet at work, and costs have not converged or declined.

For cost modeling purposes this uncertainty about the expense of the test can easily be handled by varying the model estimates of test cost. Adequate estimates of the cost of primary care office-based cervical cancer screening and resulting evaluation and treatment relevant to the US health care system are available.[32] However, the remaining parameters required for cost-effectiveness analysis hinder such estimates. Factors such as repeatability of HPV test results, incidence and types of HPV by age category, rates of progression and regression of HPV, and sensitivity and specificity for predicting outcomes over varied intervals of time are inadequately defined for using HPV tests in screening.

No recent literature specific to the United States is available to model and compare effectiveness or costs of screening or triage uses of HPV testing that have been proposed. The most complete contemporary model to date, constructed based on the United Kingdom screening system, considered adding HPV to Pap testing, replacing Pap testing with HPV testing, and adding HPV testing to surveillance of low-grade lesions versus continuing conventional surveillance. The authors concluded: "The uncertainty, as expressed by the differences between models, is so large, the results are inconclusive. Adding HPV testing to cervical cancer screening may or may not improve the (cost) effectiveness of screening. There are relatively few longitudinal HPV screening studies with enough time lapse between measurement points to

Chapter III. Results

decrease uncertainty."(p. 115)[70] When such models are generated they are further challenged by the lack of existing outcomes data even on a small scale to support or falsify the overall model or its components.

Overall, the quality of this literature is adequate for assessing the performance characteristics of the newer HPV testing methods. Hybrid Capture II and consensus PCR have good sensitivity and negative predictive value. This suggests a strong potential for a role in determining screening intervals and in triaging patients with abnormal cytology, especially among older women who have more stable HPV profiles. However, the quality of the literature is limited for the purposes of making decisions about implementing HPV testing in the US population for general screening or triage use. No published randomized trials or prospective comparisons in suitable populations were identified. In the absence of studies that relate the screening tools used to outcomes, the linkages between comparative test performance of HPV and cytologic screening tools are insufficient to judge the implications of preferentially using one combination of screening tools over another.

IV. Discussion

Context

To place choices about cervical screening in context, we need to highlight 4 factors not explicitly investigated in this systematic evidence review (SER).

First, the successes of cervical cancer screening to date, using conventional cytology, have been achieved using a diagnostic test that in other settings would be considered weak. Synthesis of the highest quality literature evaluating traditional Pap testing produces estimates that its sensitivity is 51 percent and specificity is 98 percent.[32,50]

Second, these test characteristics have proven adequate because cervical dysplasia is a slowly progressing and often self-resolving condition. The estimated progression time from severe dysplasia or carcinoma in situ to invasive cancer is 10 to 15 years.[21,32] Thus, repeated screening builds in redundant opportunities to detect abnormalities.

Third, cervical dysplasia itself does not cause morbidity or mortality. Low-grade squamous intraepithelial lesions are the most common changes detected by cytology in regularly screened populations, but they are also the most likely to resolve spontaneously. These low-grade changes are most common among young women, as are intermittent, recurrent, and resolving infection with human papilloma virus (HPV).

Fourth, an ideal screening system will optimize detection of high-grade lesions and minimize evaluation of low-grade and false-positive test results. However, this ideal can be achieved only with improved use of screening among the currently unscreened and with reliable systems for assuring follow-up of abnormal screening findings.

Chapter IV. Discussion

Major Findings and Limitations of the Literature

Who Should be Screened and How Often?

We focused this question on screening among women age 65 and older and for those who have had a hysterectomy and on identifying research that directly compared methods of selecting screening interval. In summary, all the available evidence is observational, predominantly from large population-based data sources and from a small number of prospective cohort studies. Given these sources of information, the findings of these studies are highly coherent and support the following conclusions:

- The risks of high-grade cervical lesions and cancer fall with age.

- A history of prior normal Pap tests further reduces risk.

- If screening recommendations are not modified with age, older women are disproportionately likely to have evaluations for false-positive findings.

Among previously screened women with a history of normal Pap tests, fewer than 1 individual per 1,000 screened (in some scenarios as few as 1 per 10,000) screened will have a high-grade cytologic abnormality. As an example, if the sensitivity of cytology is 60 percent and the specificity is 98 percent for detection of high-grade abnormalities, then 34 women will be evaluated for high-grade squamous intraepithelial lesion for each true high-grade cervical lesion identified; moreover, two high-grade lesions will have been missed by cytology for every three cases identified. The ratio of true positives to false positives is much higher if low-grade cytologic changes are considered. In unpublished work, Sawaya and colleagues' report that 231 additional procedures — 112 extra Pap tests, 33 colposcopies, 30 biopsies, 35 endocervical curettages, 8 endometrial biopsies, 4 dilation and curettages, 7 loop electrosurgical excision procedures, and 2 cone biopsies — were done in response to 110 Pap tests reported as atypical

Chapter IV. Discussion

squamous cells of unknown significance (ASCUS) or greater on cytology among a group of menopausal women from a prospective cohort.[44] At the conclusion of these evaluations, a case of cervical intraepithelial neoplasia 2 and a case of vaginal intraepithelial neoplasia III (not the target of cervical cancer screening) had been identified.

The literature provides fairly reliable estimates of the number of women who need to be screened to detect serious lesions. Recommendations can be made to discontinue or substantially lengthen the interval, beyond 3 years, for screening among women age 65 and older who have a history of prior normal Pap tests, depending on tolerance for missing rare cases that would have been detectable under different screening systems. The difficult trade-off between overscreening and missing rare but potentially preventable cases is a challenge for policy in this area. It suggests at minimum that women and their providers should be fully informed about the relatively larger risk of overintervention compared to the much smaller risk of failing to detect a high-grade lesion that would lead to morbidity or mortality.

Prior recommendations of the US Preventive Services Task Force to discontinue Pap testing after hysterectomy for benign disease are clearly supported and should be re-emphasized. Lastly, no direct comparisons between outcomes of proposed screening systems or interval for screening were identified. These findings confirm that this literature is substantially less well developed than that of other areas of screening tests, such as colorectal cancer screening, in which trials comparing use of screening methods over time are available to inform decision-making.

Chapter IV. Discussion

New Methods for Preparing or Evaluating Cervical Cytology

This portion of our SER evaluated the performance of new technologies for preparing or interpreting the cytology specimens; these included liquid-based cytology collection systems, neural-net screening and rescreening tools, and computer algorithms for selecting slides for screening or rescreening. This literature is fundamentally limited by lack of histologically confirmed performance measures—no gold standard is used for comparisons. Of the very few studies using colposcopy and histology to verify the diagnostic test characteristics of new tools, the most common shortcoming was failure to apply the gold standard to test negative population or a subset to allow estimation of specificity.

To receive approval from the Food and Drug Administration, each of the new technologies currently in use (ThinPrep®, PapNet®, and AutoPap®) had to demonstrate improved sensitivity over conventional Pap testing. As noted in the introduction of this chapter, improving sensitivity, especially if it amplifies detection of low-grade lesions, may not be a benefit at a population level in terms of reducing the burden of morbidity and mortality associated with cervical cancer. This is especially true if increased sensitivity is accompanied by decreased specificity, requiring evaluation of a greater number of false positives and increasing costs. It is precisely these parameters that are most poorly measured in this literature.

Overall, the quality of this literature is poor for the purposes of making decisions about choice of screening systems in US populations. No randomized trials or prospective cohort studies relate use of a screening modality over time to outcomes for individual women. The cost-effectiveness of use of new technologies has only been estimated, not measured directly. The most sophisticated and thorough cost model to date is significantly hindered by the limitations of the literature. Nonetheless, it demonstrates that, at present, new technologies are

Chapter IV. Discussion

more costly than conventional cytology; only if used in screening intervals of 3 years or longer will new technologies fall within the traditional range considered to be cost-effective ($50,000 per life-year).[32]

The Role of HPV Testing in Cervical Cancer Screening and Triage

HPV testing will have the greatest utility if it can aid in identification of those with high-grade cervical lesions that require prompt treatment *and* confirm low-risk status among those with comparatively minor Pap test abnormalities such as ASCUS or cervical intraepithelial neoplasm (CIN) 1. The performance characteristics of tests for high-risk HPV types suggest greatest benefit will emerge from the latter role or from combinations with other screening modalities in the former role to improve detection. The current literature on diagnostic test performance is of fair quality with good use of histologic tools to verify HPV test results; its primary limitation is lack of prospective and experimental evidence for its role in screening or triage.

At least 8 studies evaluating HPV testing in large populations are under way or recently completed but not yet in the published literature.[21] The Atypical Squamous Cells of Undetermined Significance/Low-Grade Squamous Intraepithelial Lesions Triage Study (ALTS) in the United States has completed enrollment of 5,060 women with ASCUS diagnosed by conventional cytology and then randomized at enrollment to immediate colposcopy, HPV testing, or repeat Pap with ThinPrep®, the latter two arms are used to triage patients to colposcopy or less intensive follow-up. Participants are followed for clinical outcomes for two years from enrollment. The ThinPrep® arm was closed before completion of the trial, suggesting that HPV is at least more effective than repeat cytology using a more sensitive cytology tool.

Chapter IV. Discussion

Definitive evidence on which to base recommendations and further model performance of HPV will be available from US and European trials within the year.

Benefits and Harms

The tangible benefit of cervical cancer screening is reduction of cervical cancer morbidity and mortality by identification of treatable precursors or early stage disease. Intangible benefits (such as reassurance for those with normal Pap tests) or concurrent preventive care at the time of Pap testing (such as screening for sexually transmitted disease or contraceptive counseling) are not documented. These intangible benefits are of particular interest because presumably they play a role in the persistence of annual screening among low-risk patients. Although some observers postulate that such overscreening occurs as a result of provider habit and the tradition of the annual pelvic examination, published evidence does not address either patient or provider knowledge, attitudes, or desire for annual testing.

Harms of screening are poorly understood. The majority of relevant research documents the psychological distress associated with the having an abnormal Pap test. Qualitative studies suggest: (1) women have unmet information needs about the meaning of abnormal results and lack factual information about what to expect during subsequent evaluation; (2) they experience distress and anxiety before and after evaluation while the diagnosis is unknown; and (3) patient education interventions can successfully reduce the fear and uncertainty associated with follow-up care for abnormal Pap testing.[74-80] Long-term concerns of women with and without definitive diagnosis of cervical abnormality and the influence of previous evaluation and outcomes on future screening behavior are unclear.

Chapter IV. Discussion

Future Research Needs

Future research must address outcomes of specific screening strategies, not only descriptive statistics and evaluation of test characteristics. On-going projects in the area of HPV testing may be the only areas of experimental research testing explicit hypotheses about the best approach to Pap testing. Until large randomized trials or well-designed prospective studies (capable of measuring and adjusting for anticipated biases) are done, especially ones that address topics such as age to discontinue screening and actual costs of new technologies, the relative benefits of one screening strategy over another are unproven and difficult to promote. The notable exception to indecision in the face of insufficient evidence is consumer demand for use of new technologies, at intervals unlikely to be cost-effective.

The community of cervical cancer researchers oriented towards preventive interventions and screening have consistently called for research of these types:

- Study of the factors that determine uptake and continuance of screening, provider and patient preferences, and adherence to appropriate screening, follow-up, and evaluation;

- Investigation of the potential for automated screening processes, such as computerized re-screening and screening, to reduce between laboratory variations in quality;

- Comparison of cytology technologies and HPV testing methods with a histologic reference standard including verification of the status of individuals, or a sample of individuals, with normal test results;

- Direct, prospective comparisons of screening strategies that include assessment of health outcomes and cost;

Chapter IV. Discussion

- Continued study of the natural history of cervical dysplasia and HPV infection;

- Investigation of the consequences of diagnosis, evaluation, and treatment of low-grade cervical intraepithelial abnormalities or detection of HPV with respect to potential psychological distress, sexual behavior, future reproductive and sexual function, and demands on the health care system;

- Trials of preventive interventions such as HPV vaccination, use of topical retinoids, and smoking cessation intervention among women with CIN;

- Outcomes studies evaluating use of HPV testing to guide cessation of Pap testing among older women, and

- Evaluation of health care systems methods for documenting the prior Pap test history of individual women, to promote appropriate screening and use of screening resources.

Advances in detecting high-grade lesions and minimizing overscreening must go hand-in-hand with active research on promoting uptake and continuance of appropriate Pap testing. The majority of cervical cancers still occur among women who are unscreened, who are inadequately screened, and who do not receive appropriate follow-up for abnormal screening test results. As Table 5 summarized in Chapter 3, a robust literature, including randomized intervention trials, provides insight into methods to promote uptake and continuance and to improve follow-up. New approaches to screening and new tools for screening need to be deployed within a care system that improves documentation of women's screening status, enhances the knowledge of both patients and providers, and optimizes reaching the unreached.

References

1. Ries LAG, Eisner MP, Kosary CL, et al., Editors. SEER Cancer Statistics Review 1973-1997. Bethesda, MD: National Cancer Institute; 2000.

2. International Agency for Research on Cancer (IARC) Working Group on the Evaluation of Cervical Cancer Screening Programmes. Screening for squamous cervical cancer: duration of low risk after negative results of cervical cytology and its implication for screening policies. *Br Med J.* 1986;293:659-664.

3. Sasieni PD, Cuzick J, Lynch-Farmery E. Estimating the efficacy of screening by auditing smear histories of women with and without cervical cancer. The National Co-ordinating Network for Cervical Screening Working Group. *Br J Cancer.* 1996;73:1001-1005.

4. United States Preventive Services Task Force. *Guide to Clinical Preventive Services.* 2nd ed. Alexandria, VA: International Medical Publishing, Inc.; 1996.

5. Landis SH, Murray T, Bolen S, Wingo PA. Cancer Statistics, 1999. *CA- Cancer J Clin.* 1999;49:8-31.

6. Daly MB, Bookman MA, Lerman CE; Female reproductive tract: cervix, endometrium, ovary. Greenwald P, Kramer BS, Weed DL, Editors. *Cancer Prevention and Control.* New York, NY: Marcel Dekker; 1995.

7. . *Healthy People 2000 Review.* Washington, DC: U.S. Dept. of Health and Human Services, Public Health Service, Centers for Disease Control and Prevention, National Center for Health Statistics; 1995.

8. Schottenfeld D, Fraumeni JFe. *Cancer Epidemiology and Prevention.* 2nd ed. New York: Oxford University Press; 1996.

9. de Vet HCW, Sturmans F, Knipschild PG. The role of cigarette smoking in the etiology of cervical dysplasia. *Epidemiology.* 1994;5(6):631-633.

10. Winkelstein W. Smoking and cervical cancer -- current status: a review. *Am J Epidemiol.* 1990;131:945-957.

11. Lyon JL, Gardner JW, West DW, Stanish WM, Hebertson RM. Smoking and carcinoma in situ of the uterine cervix. *Am J Public Health.* 1983;73:558-562.

12. Magnusson PE, Sparen P, Gyllensten U. Genetic link to cervical tumours. *Nature.* 1999;400:29-30.

13. International Agency for Research on Cancer (IARC). Monograph on the evaluation of carcinogenic risk to humans, vol. 64. Lyons: IARC; 199564.

14. Bosch FX, Manos MM, Munoz N, et al. Prevalence of human papillomavirus in cervical

References

cancer: a worldwide perspective. International biological study on cervical cancer (IBSCC) Study Group. *J Natl Cancer Inst.* 1995;87(11):796-802.

15. Jacobs MV, Snijders PJ, van den Brule AJ, Helmerhorst TJ, Meijer CJ, Walboomers JM. A general primer GP5+/GP6(+)-mediated PCR-enzyme immunoassay method for rapid detection of 14 high-risk and 6 low-risk human papillomavirus genotypes in cervical scrapings. *J Clin Microbiol.* 1997;35:791-795.

16. Burk RD, Ho GY, Beardsley L, Lempa M, Peters M, Bierman R. Sexual behavior and partner characteristics are the predominant risk factors for genital human papillomavirus infection in young women. *J Infect Dis.* 1996;174(4):679-689.

17. Dillner J, Kallings I, Brihmer C, et al. Seropositivities to human papillomavirus types 16, 18, or 33 capsids and to Chlamydia trachomatis are markers of sexual behavior. *J Infect Dis.* 1996;173(6):1394-1398.

18. Schiffman MH, Brinton LA. The epidemiology of cervical carcinogenesis. *Cancer.* 1995;76:1888-1901.

19. Meijer CJ, Helmerhorst TJ, Rozendaal L, van der Linden JC, Voorhorst FJ, Walboomers JM. HPV typing and testing in gynaecological pathology: has the time come? *Histopathology.* 1998;33:83-86.

20. Rosenthal DL, Acosta D, Peters RK. Computer-assisted rescreening of clinically important false negative cervical smears using the PAPNET Testing System. *Acta Cytologica.* 1996;40:120-126.

21. Cuzick J, Sasieni P, Davies P, et al. Systematic Review of the Role of Human Papillomavirus Testing Within a Cervical Cancer Screening Program. *Health Technol Assess.* 1999;3:I-IV and 1-196.

22. Park T, Fujiwara H, Wright TC. Molecular biology of cervical cancer and its precursors. *Cancer.* 1995;76:1902-1913.

23. Koutsky L. Epidemiology of genital human papillomavirus infection. *Am J Med.* 1997;102:3-8.

24. Hildesheim A, Hadjimichael O, Schwartz PE, et al. Risk factors for rapid-onset cervical cancer. *Am J Obstet Gynecol.* 1999;180:571-577.

25. Janerich DT, Hadjimichael O, Schwartz PE, et al. The screening histories of women with invasive cervical cancer, Connecticut. *Am J Public Health.* 1995;85:791-794.

26. Sawaya GF, Kerlikowske K, Lee NC, Gildengorin G, Washington AE. Frequency of cervical smear abnormalities within 3 years of normal cytology. *Obstet Gynecol.* 2000;96:219-223.

27. Martin-Hirsch P, Lilford R, Jarvis G, Kitchener HC. Efficacy of cervical-smear collection

References

devices: a systematic review and meta-analysis [published erratum appears in Lancet 2000 Jan 29;355(9201):414]. *Lancet.* 1999;354:1763-1770.

28. The Bethesda System for reporting cervical/vaginal cytologic diagnoses: revised after the second National Cancer Institute Workshop, April 29-30, 1991. *Acta Cytologica.* 1993;37(2):115-124.

29. Richart RM. Cervical intraepithelial neoplasia. *Pathol Annu.* 1973;8:301-328.

30. Reagan JW, Fu YS. Histologic types and prognosis of cancers of the uterine cervix. *Int J Radiat Oncol Biol Phys.* 1979;5:1015-1020.

31. Nyirjesy I. Atypical or suspicious cervical smears. An aggressive diagnostic approach. *J Am Med Assoc.* 1972;222:691-693.

32. McCrory DC, Mather DB, Bastian L. *Evaluation of Cervical Cytology: Evidence Report Number 5, Summary*Agency for Health Care Policy and Research. Rockville, MD: Agency for Health Care Policy and Research; 1999.

33. Harris RP, Helfand M, Woolf SH, et al. Current Methods of the US Preventive Services Task Force: A Review of the Process. *Am J Prev Med.* 2001;2 (3S):21-35.

34. Kainz C, Gitsch G, Heinzl H, Breitenecker G. Incidence of cervical smears indicating dysplasia among Austrian women during the 1980s. *Br J Obstet Gynaecol.* 1995;102:541-544.

35. Gram IT, Macaluso M, Stalsberg H. Incidence of cervical intraepithelial neoplasia grade III, and cancer of the cervix uteri following a negative Pap-smear in an opportunistic screening. *Acta Obstet Gynecol Scand.* 1998;77:228-232.

36. Sawaya GF, Kerlikowske K, Lee NC. When can cervical cancer screening intervals be lengthened? Outcomes following 1, 2, and 3 or more normal cervical smears. (unpublished data).

37. Forsmo S, Jacobsen BK, Stalsberg H. Cervical neoplasia in pap smears: risk of cervical intra-epithelial neoplasia (CIN) after negative or no prior smears in a population without a mass screening programme. *Int J Epidemiol.* 1996;25:53-58.

38. Gustafsson L, Sparen P, Gustafsson M, et al. Low efficiency of cytologic screening for cancer in situ of the cervix in older women. *Int J Cancer.* 1995;63:804-809.

39. Cruickshank ME, Angus V, Kelly M, McPhee S, Kitchener HC. The case for stopping cervical screening at age 50. *Brit J Obstet Gynaecol.* 1997;104:586-589.

40. Lawson HW, Lee NC, Thames SF, Henson R, Miller DS. Cervical cancer screening among low-income women: results of a national screening program, 1991-1995. *Obstet Gynecol.* 1998;92:745-752.

References

41. Cecchini S, Ciatto S, Zappa M, Biggeri A. Trends in the prevalence of cervical intraepithelial neoplasia grade 3 in the district of Florence, Italy. *Tumori.* 1995;81:330-333.

42. Mitchell HS, Giles GG. Cancer diagnosis after a report of negative cervical cytology [see comments]. *Med J Aust.* 1996;164:270-273.

43. Sigurdsson K. Trends in cervical intra-epithelial neoplasia in Iceland through 1995: evaluation of targeted age groups and screening intervals. *Acta Obstet Gynecol Scand.* 1999;78:486-492.

44. Sawaya GF, Grady D, Kerlikowske K, et al. The positive predictive value of cervical smears in previously-screened postmenopausal women: the Heart and Estrogen/progestin Replacement Study (HERS). *Ann Intern Med.* 2000;133:942-950.

45. Miller AB, Visentin T, Howe GR. The effect of hysterectomies and screening on mortality from cancer of the uterus in Canada. *Int J Cancer.* 1981;27:651-657.

46. Eaker ED, Vierkant RA, Konitzer KA, Remington PL. Cervical cancer screening among women with and without hysterectomies. *Obstet Gynecol.* 1998;91:551-555.

47. Fox J, Remington P, Layde P, Klein G. The effect of hysterectomy on the risk of an abnormal screening Papanicolaou test result. *Am J Obstet Gynecol.* 1999;180:1104-1109.

48. Pearce K, Haefner H, Sarwar S, Nolan T. Cytopathological findings on vaginal Papanicolaou smears after hysterectomy for benign gynelogical disease. *N Engl J Med.* 1996;335:1559-1562.

49. Weintraub J, Morabia A. Efficacy of a liquid-based thin layer method for cervical cancer screening in a population with a low incidence of cervical cancer. *Diagn Cytopathology.* 2000;22:52-59.

50. Nanda K, McCrory DC, Myers ER, et al. Accuracy of the Papanicolaou test in screening for and follow-up of cervical cytologic abnormalities: a systematic review. *Ann Intern Med.* 2000;132:810-819.

51. Hutchinson ML, Zahniser DJ, Sherman ME, et al. Utility of liquid-based cytology for cervical carcinoma screening: results of a population-based study conducted in a region of Costa Rica with a high incidence of cervical carcinoma. *Cancer.* 1999;87:48-55.

52. Roberts JM, Gurley AM, Thurloe JK, Bowditch R, Laverty CRA. Evaluation of the ThinPrep Pap test as an adjunct to the conventional Pap smear. *Med J Aust.* 1997;167:466-469.

53. Bolick DR, Hellman DJ. Laboratory implementation and efficacy assessment of the

References

ThinPrep cervical cancer screening system. *Acta Cytologica.* 1998;42:209-213.

54. Chock C, Irwig L, Berry G, Glasziou P. Comparing dichotomous screening tests when individuals negative on both tests are not verified. *J Clin Epidemiol.* 1997;50:1211-1217.

55. Kaufman RH, Schreiber K, Carter T. Analysis of atypical squamous (glandular) cells of undetermined significance smears by neural network-directed review. *Obstet Gynecol.* 1998;91:556-560.

56. Jenny J, Isenegger I, Boon ME, Husain OA. Consistency of a double PAPNET scan of cervical smears. *Acta Cytologica.* 1997;41:82-87.

57. Kok MR, Boon ME, Schreiner-Kok PG, Koss LG. Cytological recognition of invasive squamous cancer of the uterine cervix: comparison of conventional light-microscopical screening and neural network-based screening. *Hum Pathol.* 2000;31:23-28.

58. Wilbur DC, Bonfiglio TA, Rutkowski MA, et al. Sensitivity of the AutoPap 300 QC System for cervical cytologic abnormalities. Biopsy data confirmation. *Acta Cytol.* 1996;40:127-132.

59. Wilbur DC, Prey MU, Miller WM, Pawlick GF, Colgan TJ, Dax Taylor D. Detection of high grade squamous intraepithelial lesions and tumors using the AutoPap System: results of a primary screening clinical trial. *Cancer.* 1999;87:354-358.

60. Duggan MA. Papnet-assisted, primary screening of cervico-vaginal smears. *Eur J Gynaecol Oncol.* 2000;21:35-42.

61. Petry KU, Bohmer G, Iftner T, Flemming P, Stoll M, Schmidt RE. Human papillomavirus testing in primary screening for cervical cancer of human immunodeficiency virus-infected women, 1990-1998. *Gynecol Oncol.* 1999;75:427-431.

62. Womack SD, Chirenje ZM, Blumenthal PD, et al. Evaluation of a human papillomavirus assay in cervical screening in Zimbabwe. *Br J Obstetr Gynaecol.* 2000;107:33-38.

63. Womack SD, Chirenje ZM, Gaffikin L, et al. HPV-based cervical cancer screening in a population at high risk for HIV infection. *Int J Cancer.* 2000;85:206-210.

64. Wright TCJ, Denny L, Kuhn L, Pollack A, Lorincz A. HPV DNA testing of self-collected vaginal samples compared with cytologic screening to detect cervical cancer. *JAMA.* 2000;283:81-86.

65. Schiffman M, Herrero R, Hildesheim A, et al. HPV DNA testing in cervical cancer screening: results from women in a high-risk province of Costa Rica. *J Am Med Assoc.* 2000;283:87-93.

References

66. Bollen LJ, Tjong-A-Hung SP, van der Velden J, et al. Human papillomavirus deoxyribonucleic acid detection in mildly or moderately dysplastic smears: a possible method for selecting patients for colposcopy. *Am J Obstet Gynecol.* 1997;177:548-553.

67. Sigurdsson K, Arnadottir T, Snorradottir M, Benediktsdottir K, Saemundsson H. Human papillomavirus (HPV) in an Icelandic population: the role of HPV DNA testing based on hybrid capture and PCR assays among women with screen-detected abnormal pap smears. *Int J Cancer.* 1997;72:446-452.

68. Manos MM, Kinney WK, Hurley LB, et al. Identifying women with cervical neoplasia: using human papillomavirus DNA testing for equivocal Papanicolaou results. *J Am Med Assoc.* 1999;281:1605-1610.

69. Herrington CS, Evans MF, Hallam NFeal. Human papillomavirus status in the prediction of high-grade cervical intra-epithelial neoplasia in patient with persistent low-grade cervical cytological abnormalities. *Br J Cancer.* 1995;71:206-209.

70. Cuzick J, Sasieni P, Davies P, et al. A systematic review of the role of human papillomavirus testing within a cervical screening programme. *Health Technol Assess.* 1999;3:i-iv, 1-196.

71. Adam E, Kaufman RH, Berkova Z, Icenogle J, Reeves WC. Is human papillomavirus testing an effective triage method for detection of high-grade (grade 2 or 3) cervical intraepithelial neoplasia? *Am J Obstet Gynecol.* 1998;178:1235-1244.

72. Hillemanns P, Kimmig R, Huttemann U, Dannecker C, Thaler CJ. Screening for cervical neoplasia by self-assessment for human papillomavirus DNA. *Lancet.* 1999;354:1970.

73. Sun XW, Ferenczy A, Johnson D, et al. Gynecology: Evaluation of the Hybrid Capture human papillomavirus deoxyribonucleic acid detection test. *Am J Obstet Gynecol.* 1995;173:1432-1437.

74. Kavanagh AM, Broom DH. Women's understanding of abnormal cervical smear test results: a qualitative interview study. *Br Med J.* 1997;1388-1391.

75. Lauver D, Rubin M. Women's concerns about abnormal Papanicolaou test results. *JOGNN.* 1990;20:154-159.

76. Barsevick AM, Lauver D. Women's informational needs about coploscopy. *IMAGE: J Nurs Scholarship.* 1990;22:23-26.

77. Stewart DE, Lickrish GM, Sierra S, Parkin H. The effect of educational brochures on knowledge and emotional distress in women with abnormal Papanicolaou smears. *Obstet Gynecol.* 1993;81:280-282.

78. Somerset M, Peters TJ. Intervening to reduce anxiety for women with mild dyskaryosis:

References

do we know what works and why? *J Adv Nurs.* 1998;28:563-570.

79. Somerset M, Baxter K, Wilkinson C, Peters TJ. Mildy dyskaryotic smear results: does it matter what women know? *Int J Fam Pract.* 1998;15:537-542.

80. Schofield MJ, Sanson-Fisher R, Halpin S, Redman S. Notification and follow-up of Pap test results: current practice and women's preferences. *Prev Med.* 1994;23:276-283.

81. Gustafsson L, Sparen P, Gustafsson M, Wilander E, Bergstrom R, Adami HO. Efficiency of organised and opportunistic cytological screening for cancer in situ of the cervix. *Br J Cancer.* 1995;72:498-505.

82. Mitchell H, Medley G, Higgins V. An audit of the women who died during 1994 from cancer of the cervix in Victoria, Australia. *Aust N Z J Obstet Gynaecol.* 1996;36:73-76.

83. Colgan TJ, Patten SFJr, Lee JS. A clinical trial of the AutoPap 300 QC system for quality control of cervicovaginal cytology in the clinical laboratory. *Acta Cytol.* 1995;39:1191-1198.

84. Patten SFJ, Lee JS, Wilbur DC, et al. The AutoPap 300 QC System multicenter clinical trials for use in quality control rescreening of cervical smears: II. Prospective and archival sensitivity studies. *Cancer.* 1997;81:343-347.

85. Lee JS, Kuan L, Oh S, Patten FW, Wilbur DC. A feasibility study of the AutoPap system location-guided screening. *Acta Cytologica.* 1998;42:221-226.

86. Stevens MW, Milne AJ, James KA, Brancheau D, Ellison D, Kuan L. Effectiveness of automated cervical cytology rescreening using the AutoPap 300 QC System. *Diagn Cytopathol.* 1997;16:505-512.

87. Wilbur DC, Bonfiglio TA, Rutkowski MA, et al. Sensitivity of the AutoPap 300 QC System for cervical cytologic abnormalities. Biopsy data confirmation. *Acta Cytol.* 1996;40:127-132.

88. Wilbur DC, Prey MU, Miller WM, Pawlick GF, Colgan TJ. The AutoPap system for primary screening in cervical cytology. Comparing the results of a prospective, intended-use study with routine manual practice. *Acta Cytologica.* 1998;42:214-220.

89. Wilbur DC, Prey MU, Miller WM, Pawlick GF, Colgan TJ, Dax Taylor D. Detection of high grade squamous intraepithelial lesions and tumors using the AutoPap System: results of a primary screening clinical trial. *Cancer.* 1999;87:354-358.

90. Ashfaq R, Liang Y, Saboorian MH. Evaluation of PAPNET system for rescreening of negative cervical smears. *Diagn Cytopathol.* 1995;13:21-36.

References

91. Slagel DD, Zaleski S, Cohen MB. Efficacy of automated cervical cytology screening. *Diagn Cytopathol.* 1995;13:26-30.

92. Farnsworth A, Chambers FM, Goldschmidt CS. Evaluation of the PAPNET system in a general pathology service. *Med J Aust.* 1996;165:429-431.

93. Ashfaq R, Saliger F, Solares B, et al. Evaluation of the PAPNET system for prescreening triage of cervicovaginal smears. *Acta Cytol.* 1997;41:1058-1064.

94. Duggan MA, Brasher P. Paired comparison of manual and automated Pap test screening using the PAPNET system. *Diagn Cytopathol.* 1997;17:248-254.

95. Halford JA, Wright RG, Ditchmen EJ. Quality assurance in cervical cytology screening. Comparison of rapid rescreening and the PAPNET Testing System. *Acta Cytol.* 1997;41:79-81.

96. Jenny J, Isenegger I, Boon ME, Husain OA. Consistency of a double PAPNET scan of cervical smears. *Acta Cytol.* 1997;41:82-87.

97. Kaufman RH, Schreiber K, Carter T. Analysis of atypical squamous (glandular) cells of undetermined significance smears by neural network-directed review. *Obstet Gynecol.* 1998;91:556-560.

98. Mango LJ, Valente PT. Neural-network-assisted analysis and microscopic rescreening in presumed negative cervical cytologic smears. A comparison. *Acta Cytologica.* 1998;42:227-232.

99. Mitchell H, Medley G. Detection of unsuspected abnormalities by PAPNET-assisted review. *Acta Cytologica.* 1998;42:260-264.

100. O'Leary TJ, Tellado M, Buckner SB, Ali IS, Stevens A, Ollayos CW. PAPNET-assisted rescreening of cervical smears: cost and accuracy compared with a 100% manual rescreening strategy. *J Am Med Assoc.* 1998;279:235-237.

101. Kok MR, Boon ME, Schreiner-Kok PG, Koss LG. Cytological recognition of invasive squamous cancer of the uterine cervix: comparison of conventional light-microscopical screening and neural network-based screening. *Hum Pathol.* 2000;31:23-28.

102. Corkill M, Knapp D, Martin J, Hutchinson ML. Specimen adequacy of ThinPrep sample preparations in a direct-to-vial study. *Acta Cytologica.* 1997;41:39-44.

103. Lee KR, Ashfaq R, Birdsong GG, Corkill ME, McIntosh KM, Inhorn SL. Comparison of conventional Papanicolaou smears and a fluid-based, thin-layer system for cervical cancer screening. *Obstet Gynecol.* 1997;90:278-284.

104. Roberts JM, Gurley AM, Thurloe JK, Bowditch R, Laverty CR. Evaluation of the ThinPrep Pap test as an adjunct to the conventional Pap smear. *Med J Aust.*

References

1997;167:466-469.

105. Bolick DR, Hellman DJ. Laboratory implementation and efficacy assessment of the ThinPrep cervical cancer screening system. *Acta Cytol.* 1998;42:209-213.

106. Inhorn SL, Linder J, Wilbur DC, Zahniser D. Diagnosis of invasive cervical carcinoma with fluid-based, thin-layer slide preparation method. *Prim. Care Update Ob Gyns.* 1998;5:164.

107. Papillo JL, Zarka MA, St John TL. Evaluation of the ThinPrep Pap test in clinical practice. A seven-month, 16,314-case experience in northern Vermont. *Acta Cytologica.* 1998;42:203-208.

108. Sherman ME, Mendoza M, Lee KR, et al. Performance of liquid-based, thin-layer cervical cytology: correlation with reference diagnoses and human papillomavirus testing. *Modern Pathol.* 1998;11:837-843.

109. Hutchinson ML, Zahniser DJ, Sherman ME, et al. Utility of liquid-based cytology for cervical carcinoma screening: results of a population-based study conducted in a region of Costa Rica with a high incidence of cervical carcinoma. *Cancer.* 1999;87:48-55.

Appendix A
Acknowledgments

Appendix A. Acknowledgments

Appendix A
Acknowledgments

This study was supported by Contract 290-97-0011 from the Agency for Healthcare Research and Quality (Task No. 3 to support the US Preventive Services Task Force). We acknowledge the ongoing guidance of David Atkins, M.D., M.P.H., Director of the Preventive Services Program at AHRQ, and the continuing support of Jacqueline Besteman, J.D., M.A., the AHRQ Task Order Officer for this project. The investigators deeply appreciate the considerable support and contributions of faculty and staff from the University of North Carolina at Chapel Hill – Russ Harris, M.D., M.P.H., Michael Pignone, M.D., M.P.H., Lynn Whitener, Dr.P.H., M.S.L.S., Barbara Starrett, M.H.A., Catherine Mills Burchell, M.A., and Scott Pohlman, M.S. We would like to express our special thanks to those who helped in so many practical ways to prepare the final document – Cheryl Roman, B.A., Christine Shia, M.A., and Melissa McPheeters, M.P.H.

We are equally grateful to Kathleen Lohr, Ph.D., Linda Lux, M.P.H., Anjolie Idicula Laubach, B.A., and Sonya Sutton, B.S.P.H., from RTI International for their continued guidance and assistance with the project and to Sheila White and Loraine Monroe for superior secretarial support. In addition, we would like to extend our thanks to the following expert peer reviewers, who provided input on an earlier version of this report: Michael L. LeFevre, M.D.,M.S.P.H., of the University of Missouri School of Medicine, Columbia, MO; Jean-Marie Moutquin, M.D., M.Sc., F.R.S.C.(C).of the Canadian Task Force on Preventive Care, Sherbrooke, Quebec; Evan Myers, M.D., M.P.H., of Duke University Medical Center, Durham, N.C.; Kenneth Noller, M.D.,

Appendix A. Acknowledgments

of the American College of Obstetricians and Gynecologists, Washington, D.C.; and Electra Paskett, Ph.D., of the Wake Forest University Baptist Medical Center, Winston-Salem, N.C.

Appendix B
Methods

Appendix B: Methods

USPSTF CERVICAL CANCER SCREENING

Article Screen & Abstract

Key Question 1A:
Who should be screened for cervical cancer and how often?

Specifically, what are the outcomes (benefits, harms, and costs) associated with screening among women age 65 and older?

Initials of Reviewer __ __ __   Date __ / __ / __   Unique Article Identifier __ __ __ __
                                     (mm,dd,yy)

SCREEN

Is this an original research article?	☐ Yes	☐ No
Does the study address screening/outcomes of screening for squamous cell carcinoma of the cervix?	☐ Yes	☐ No
Is the age distribution of the study population specified?	☐ Yes	☐ No
Are women ≥ 50 years old included?	☐ Yes	☐ No
Is stratified data about outcomes/risk/etc. provided by age?	☐ Yes	☐ No
Is the denominator provided from which the cases arose?	☐ Yes	☐ No*
*If not, are rates provided that indicate the denominator is known?	☐ Yes	☐ No

ALL ABOVE MUST BE "YES" TO INCLUDE FOR KQ3	☐ Include	☐ Reject

Other Screening Questions

Does the article include histologically verified measures of **Pap** test or **HPV** test diagnostic characteristics?	☐ Yes	☐ No
Does the article report findings for women who have had a **hysterectomy**?	☐ Yes	☐ No
Does the article report costs of screening based on actual **cost of care** in the US?	☐ Yes	☐ No
Does the article report **harms** of screening/testing?	☐ Yes	☐ No

Appendix B: Methods

USPSTF CERVICAL CANCER SCREENING

Article Screen & Abstract

Key Question 1B:
Who should be screened for cervical cancer and how often?

What are the outcomes (benefits, harms, and costs) associated with screening among women who have had a hysterectomy?

Initials of Reviewer __ __ __   Date __ / __ / __   Unique Article Identifier __ __ __ __
(mm,dd,yy)

SCREEN

Is this an original research article?	❏ Yes	❏ No
Does the study address screening/outcomes of screening for squamous cell carcinoma of the cervix?	❏ Yes	❏ No
Are outcomes of screening compared for women with and without a hysterectomy?	❏ Yes	❏ No

ALL ABOVE MUST BE "YES" TO INCLUDE FOR KQ3	❏ Include	❏ Reject

Other Screening Questions

Does the article include histologicaly verified measures of **Pap** test or **HPV** test diagnostic characteristics?	❏ Yes	❏ No
Does the article include stratification by **age**, with women age 50 or older included?	❏ Yes	❏ No
Does the article report costs of screening based on actual **cost of care** in the US?	❏ Yes	❏ No
Does the article report **harms** of screening/testing?	❏ Yes	❏ No

Appendix B: Methods

USPSTF CERVICAL CANCER SCREENING

Article Screen & Abstract

Key Question 2:
Do new methods for preparing/evaluating cervical cytology improve diagnostic yield?

Initials of Reviewer __ __ __   Date __ / __ / __   Unique Article Identifier __ __ __ __
                                        (mm,dd,yy)

SCREEN   The new method(s) under evaluation: _____

Is this an original research article?	☐ Yes	☐ No
Was Pap testing done as a screening test?	☐ Yes	☐ No
Does the study include at least 50 subjects/specimens with Pap results *and* new method testing?	☐ Yes	☐ No
Was the new method used for primary screening or as an adjunct to primary screening?	☐ Yes	☐ No
Were Pap *and* the new method compared with a reference standard (histology/colposcopy)?	☐ Yes	☐ No
Were the Pap, new test, and reference standard obtained within a three month window?	☐ Yes	☐ No
Can all cells of a 2x2 table be completed?	☐ Yes	☐ No

ALL ABOVE MUST BE "YES" TO INCLUDE FOR KQ2	☐ Include	☐ Reject

Other Screening Questions

Does the article include stratification by **age**, with women age 50 or older included?	☐ Yes	☐ No
Does the article report findings for women who have had a **hysterectomy**?	☐ Yes	☐ No
Does the article report costs of screening based on actual **cost of care** in the US?	☐ Yes	☐ No
Does the article report **harms** of screening/testing?	☐ Yes	☐ No

Appendix B: Methods

USPSTF CERVICAL CANCER SCREENING

Article Screen & Abstract

Key Question 3:
What is the role of HPV testing in cervical cancer screening strategies?

Initials of Reviewer __ __ __   Date __ / __ / __   Unique Article Identifier __ __ __ __
(mm,dd,yy)

SCREEN

Is this an original research article?	❏ Yes	❏ No
Was Pap testing done as a screening test?	❏ Yes	❏ No
Does the study include at least 50 subjects/specimens with Pap results *and* HPV testing?	❏ Yes	❏ No
Was HPV testing used for primary screening or as an adjunct to primary screening?	❏ Yes	❏ No
Were Pap *and* HPV testing compared with a reference standard (histology/colposcopy)?	❏ Yes	❏ No
Were the Pap, HPV test, and reference standard obtained within a three month window?	❏ Yes Elapsed time: ___	❏ No ❏ DK
Can all cells of a 2x2 table be completed?	❏ Yes	❏ No

ALL ABOVE MUST BE "YES" TO INCLUDE FOR KQ3	❏ Include	❏ Reject

Other Screening Questions

Does the article include stratification by **age**, with women age 50 or older included?	❏ Yes	❏ No
Does the article report findings for women who have had a **hysterectomy**?	❏ Yes	❏ No
Does the article report **harms** of screening/testing?	❏ Yes	❏ No

**Appendix C
Evidence Tables**

Appendix C. Evidence Tables

Glossary of terms

AGCUS	Atypical Glandular Cells of Unknown Significance
AmInd	American Indian
ASCUS	Atypical Squamous Cells Of Unknown Significance
CA	Cancer
CI	Confidence Interval
CIN	Cervical Intraepithelial Neoplasia
CIS	Carcinoma In Situ
CPI	Clinical Position Imaging
DNA	Deoxyribonucleic Acid
Denom	Denominator
EC	European Community
EPC	Evidence-based Practice Center
FDA	Food and Drug Administration
FN	False Negative
FPR	False Positive Rate
GS	Gold Standard
HC	Hybrid Capture
HC I	Hybrid Capture, First Generation
HC II	Hybrid Capture – Second Generation
HCT	Hybrid Capture Table
Hisp	Hispanic
HPV	Human Papilloma Virus
HSIL	High-Grade Squamous Intraepithelial Lesion

Appendix C. Evidence Tables

Hx	History
LA	Louisiana
LEEP	Loop Electrosurgical Excision Procedure
LSIL	Low-Grade Sqamous Intraephithelial Lesion
NBCS	National Birth Center Study
NPV	Negative Predictive Value
NS	Not Specified – Not Statistically Significant
OR	Odds Ratio
PCR	Polymerase Chain Reaction
PPV	Positive Predictive Value
PY®	Per year
QC	Quality Check
Ref. Std.	Reference Standard
RR	Relative Risk
SCC	Squamous Cell Carcinoma
SE	Sensitivity
SER	Systematic Evidence Review
SIL	Squamous Intraepithelial Lesion
S/P	Status Post
SP	Specificity
TPR	True Positive Rate
UICC	Union Internationale Contre le Cancer
UK	United Kingdom
UKN	Unknown

Appendix C. Evidence Tables

VAIN	Vaginal Intraepithelial Neoplasia
WI	Wisconsin
WVC	Wart Virus Changes
YOA	Years Of Age

Appendix C. Evidence Tables

Evidence Table 1A. Screening Among Older Women

Source: Author, Year	Study Design & Characteristics	Stated Objective(s)	Location & Time Period	Data Source(s)
Cecchini et al., 1995[41]	Population-based, retrospective cohort	To investigate the detection rate of CIN3 in previously unscreened women, in order to reveal trends over time in the prevalence of CIN3	Florence, Italy 1973-1992	Cervical cancer screening database; centralized colposcopy and pathology services
Gustafsson et al., 1995[81]	Population-based, retrospective cohort	To estimate the efficiency of detecting cancer in situ in cytologic screening for cervical cancer at different ages with emphasis on women older than 50	Uppsala County, Sweden 1969-1988	Cervical cancer screening and pathology database

Appendix C. Evidence Tables

Evidence Table 1A. Screening Among Older Women (cont'd)

Participants	Screening Program	Outcomes & Measures	Relevant Results	Quality Consideration
N = 287,295 Pap smears Age 20-60 at first screening or first screening within 10 years	N/A (identified cases at entry into screening)	Primary outcome = histologically proven CIN3 (n=648)	AGE: Within the age groups from 30-54 age/cohort specific rates of CIN 3 are increasing over secular time. Patterns are variable among older age groups over time. Recent rates by age: Age 20-24 1.58 cases/1000py 25-9 2.51 30-4 4.94 35-9 4.01 40-4 4.29 45-9 3.40 50-4 3.90 55-9 1.88	Denom=2 Attrition=DNA Indications=0 (DNA) Interval=0 DNA Age=2 Relevance=0.5 Standard=2 Score=6.5 Grade=Poor
N = 118,890 women with 466,259 Pap smears in registry before diagnosis of cancer in situ or invasive cancer	Not specified	Primary outcome = cancer in situ n=1076	AGE: 17% of smears from women ≥50 yoa. CIS among ≥ 50 was 4.6%. Therefore detection ratio per thousand smears of 0.63. Contrasted with peak detection ratio of 3.3 for women ages 25-35; and combined ratio for those under 50 of 2.6. AGE & SCREENING HISTORY: (Subanalysis – n = 5,893) Among women with 3 normal smears between 41 and 49, 60 (1%) had Class 3-5 smear after 50, including 4 CIS, 3 invasive cancers and 2 adenocarcinomas	Denom=2 Attrition=1 Indications=2 Interval=1 Age=2 Relevance=1 Standard=2 Score=11 Grade=Fair

Appendix C. Evidence Tables

Evidence Table 1A. Screening Among Older Women (cont'd)

Source: Author, Year	Study Design & Characteristics	Stated Objective(s)	Location & Time Period	Data Source(s)
Kainz et al., 1995[34]	Retrospective cohort	To assess the incidence of CIN detected by cervical smears among women of different ages and to compare two quintennia	Vienna, Austria 1980-1989	Computerized records from a single cytology lab. Included Pap smears were obtained by gynecologists who used only this cytology lab

Appendix C. Evidence Tables

Evidence Table 1A. Screening Among Older Women (cont'd)

Participants	Screening Program	Outcomes & Measures	Relevant Results	Quality Consideration
N = 12,604 non-pregnant women with 2 Paps within 350-380 days; negative cytology on 1st smear. Women with history of "inflammatory smears" or CIN excluded	Annual advised	Primary outcome = CIN 1-3 on cytology (Histology gold standard for quality control stated, not described)	AGE: Incidence of CIN 1-3 Age 1980-84 1985-89 RR(95%CI) ≤ 20 4/697 4/593 1.2(0.3,4.7) 21-30 12/2570 34/2102 3.5(1.9,6.5) 31-40 5/1737 11/1150 3.3(1.2, 9.0) ≥41 8/1694 2/2061 0.2(0.1,0.8)	Denom=2 Attrition=2 Indications=2 Interval=2 Age=1 Relevance=1 Standard=1+ Score=11 Grade=Fair

Appendix C. Evidence Tables

Evidence Table 1A. Screening Among Older Women (cont'd)

Source: Author, Year	Study Design & Characteristics	Stated Objective(s)	Location & Time Period	Data Source(s)
Forsmo et al., 1996[37]	Retrospective cohort	To examine…risk of cancer and CIN in Pap smears from women without previously reported positive smears	Finnmark & Troms counties, Norway 1988-1990	Computerized records from a single cytology lab

Appendix C. Evidence Tables

Evidence Table 1A. Screening Among Older Women (cont'd)

Participants	Screening Program	Outcomes & Measures	Relevant Results	Quality Considerations

N = 40,536 women with 58,271 Paps; no prior abnormal cytology including smears with evidence of HPV or Herpes virus
 	
Age 13-100, mean age: 36
	No systematic screening program in place; centralized recall for follow-up of abnormal Paps only	Primary outcome = CIN 1-cancer on cytology
		
		(No histology based data)
			AGE:
			Incidence
			
			        Cases Cases Total
			        CIN 1-2 CIN 3 N
			$\leq$ 19 43 1 4,962
			20-29 203 21 18,531
			30-39 96 27 13,761
			40-49 43 13 9,955
			50-59 22 12 5,370
			60-69 10 7 3,568
			   70 10 9 2,124
			
			ORs* CIN 3 & Cancer Among Older Women (95%CI)
			Age CIN 3 Cancer
			50-59 1.0 (ref) 1.0 (ref)
			60-69 0.6 (0.2, 1.4) 2.5 (0.7, 9.7)
			$\geq$ 70 0.8 (0.2, 0.8) 1.9 (0.4, 8.1)
			* Adjusted for time since last smear
			
			INTERVAL:
			(months) OR^ (95%CI)
			First 1.0 (ref) 1.0 (ref)
			  12 0.1 (0.0, 0.7) ---
			13-35 0.04 (0.0, 0.2) 0.04 (0.0, 0.3)
			36-59 0.12 (0.0, 0.4) 0.1 (0.1, 0.8)
			  60 0.26 (0.1, 0.7) 0.5 (0.2, 1.6)
			^ Adjusted for age
				Denom=2
				Attrition=2
				Indications=2
				Interval=2
				Age=2
				Relevance=1
				Standard=0
				Score=11
				
				Grade=Fair

Appendix C. Evidence Tables

Evidence Table 1A. Screening Among Older Women (cont'd)

Source: Author, Year	Study Design & Characteristics	Stated Objective(s)	Location & Time Period	Data Source(s)
Mitchell et al., 1996[82]	1) Prospective Cohort; 2) Case series	1) To determine the annual rate of interval squamous cancer of the cervix after a negative [Pap test]; and 2) To evaluate the proportion of women with cervical cancer who received negative reports during the three years before the cancer diagnosis	Victoria province, Australia 1990-1993	Centralized cytology registry and cancer registry
Cruickshank et al., 1997[39]	Retrospective Birth Cohort	To determine the pattern of abnormal cervical cytology in women aged 50 to 60 years and to determine whether the development of cervical neoplasia in this age group is confined to women who have been inadequately screened	Grampian region, UK 1989-1994	Centralized cytology registry

Appendix C. Evidence Tables

Evidence Table 1A. Screening Among Older Women (cont'd)

Participants	Screening Program	Outcomes & Measures	Relevant Results	Quality Considerations
1) 368,051 women with normal Pap in 1990 Age <70 2) 233 women with new diagnosis of cervical cancer in 1993 (all ages)	Recommends screening every 2 yrs after a negative Pap, stopping at 70	Primary outcome = microinvasive and invasive cervical cancer as documented by cancer registry which includes histology	AGE: Interval cancer rate: (95%CI) Age <35: 2.54/100,000 (1.5, 4.3) Ages 35-69: 2.53/100,000 (1,5, 4.3) INTERVAL: Of 167 women with diagnosis of squamous cell cancer, 34 (20%) had a negative Pap report within the preceding 3 years	Denom=2 Attrition=1 Indications=0 Interval=2 Age=0 Relevance=1 Standard=2 Score=8 Grade=Poor

N = 30,527 women 25,216 hysterectomy
 	
Born between 10/2/33 & 10/1/44 (i.e. 50-60 at time of analysis); and without hysterectomy
	Call and recall screening system to achieve screening every 3 years between ages 20-60
	
	More than 93% of population has been screened at least once
		Primary outcome = "non-negative smear" defined as borderline atypia or greater n= 229
		
		Secondary outcome = Histologic diagnoses
			AGE
			     Non-neg. N %
			     Smear Screened Non-neg
			50-53 161 10,365 1.4
			54-57 54 8,913 0.6
			58-60 14 6,469 0.2
			
			Histology among 164 women who had biopsy for non-negative smear
			Negative 27
			Viral changes only 16
			CIN 1 13
			CIN 2 19
			CIN 3 70
			Micro-invasive* 4
			Invasive cancer* 4
			Clinically detected 7
			Other 4
			* Screen detected
			
			Of 8 cancers, 1 first screened after age 50; 3 had prior known abnormal smears; 2 had their two prior smears at intervals >5 yrs; 1 between 4-5 years. None were found who had interval ≤ 3 years with normal smears
				Denom=2
				Attrition=2
				Indications=2
				Interval=2
				Age=1
				Relevance=1
				Standard=1
				Score=11
				
				Grade=Fair

Appendix C. Evidence Tables

Evidence Table 1A. Screening Among Older Women (cont'd)

Source: Author, Year	Study Design & Characteristics	Stated Objective(s)	Location & Time Period	Data Source(s)
Gram et al., 1998[35]	Prospective cohort	To investigate the influence of screening history on the diagnosis of CIN 3 and cervical cancer in an opportunistic screening program	Troms and Finnmarker counties, Norway 1980-1989	Computerized records from a single cytology lab

Appendix C. Evidence Tables

Evidence Table 1A. Screening Among Older Women (cont'd)

Participants	Screening Program	Outcomes & Measures	Relevant Results	Quality Consideration
N = 41,212 women with negative Pap at entry Born 1920-69; no CIN or cervical cancer hx; no hx of cervical biopsy	No systematic screening program in place; centralized recall for follow-up of abnormal Paps only	Primary outcome = CIN 3 or cervical cancer 396 incident in 175,673 person yrs	AGE: Incidence rate for CIN 3 and cancer was highest among women aged 25-29 and decreased with age thereafter: Age Person Years Cases 10,000 PYR 20-24 33,663 68 25-29 34,349 136 30-34 27,325 87 35-39 27,325 44 > 40 48,666 61 INTERVAL: Age-Adjusted Risk of CIN III/Cancer Years RR (95% CI) < 1 1.0 1 1.5 (1.2, 2.0) 2 2.4 (1.8, 3.3) 3+ 12.2 (9.2, 16.3)	Denom=2 Attrition=0 Indications=2 Interval=2 Age=2 Relevance=1 Standard=1 Score=10 Grade=Fair

Appendix C. Evidence Tables

Evidence Table 1A. Screening Among Older Women (cont'd)

Source: Author, Year	Study Design & Characteristics	Stated Objective(s)	Location & Time Period	Data Source(s)
Lawson et al., 1998[40]	Prospective Cohort	To evaluate the results of cervical cytology screening in the National Breast and Cervical Cancer Early Detection Program (NBCCEDP)	US (sites in 22 states and 5 native tribal programs) 10/1/91-6/30/95	Prospective data collection as part of NBCCEDP

Appendix C. Evidence Tables

Evidence Table 1A. Screening Among Older Women (cont'd)

Participants	Screening Program	Outcomes & Measures	Relevant Results	Quality Considerations
N = 312,858 women with 407,246 Paps Services intended for low-income women Age 40-49 25.2% 50-64 24.9% ≥ 65 7.6% Black, non-Hisp 10.3% Hisp 21.2% Asian 2.3% AmInd 10.4%		Primary Outcome = histologic diagnoses of CIN 2,3, CIS and invasive cancer (all cases with histology from CIN 1, up reported) Secondary outcome = cytology findings	AGE: First Screening Cycle Rates* of histology-proven diagnoses 　　　　CIN 2　CIN 3/CIS　Invasive < 30　11.0　　7.9　　　0.1 30-39　4.9　　6.7　　　0.4 40-49　2.0　　3.2　　　0.5 50-64　0.9　　1.7　　　0.6 ≥ 65　0.5　　1.4　　　0.3 Second Screening Cycle Rates* of histology-proven diagnoses 　　　　CIN 2　CIN 3/CIS　Invasive < 30　7.6　　4.2　　　0.1 30-39　3.8　　4.1　　　0.4 40-49　1.7　　1.3　　　0.2 50-64　0.7　　1.2　　　0.1 ≥ 65　0.8　　0.7　　　0.0 * Rates = diagnoses per 1000 Pap smears 　　　　First test　　Subsequent Age　　% Abnormal　%Abnormal < 30　　8.2　　　　7.4 30-39　4.3　　　　4.2 40-49　2.3　　　　2.3 50-64　1.4　　　　1.5 ≥ 65　1.0　　　　1.0	Denom=2 Attrition=0 Indications=1 Interval=0 Age=2 Relevance=2 Standard=0 Score=7 Grade=Fair

Appendix C. Evidence Tables

Evidence Table 1A. Screening Among Older Women (cont'd)

Source: Author, Year	Study Design & Characteristics	Stated Objective(s)	Location & Time Period	Data Source(s)
Sigurdson et al., 1999[43]	Retrospective Cohort	To evaluate the UICC and EC recommendations regarding target age group and screening interval	Iceland 1966-1995	Records of the national cervical cancer screening program
Sawaya et al., 2000[26]	Prospective cohort	To compare cervical screening outcomes associated with age and screening intervals: 1, 2, and 3 years	US 1991-1998	Prospective data collection as part of National Breast & Cervical Cancer Early Detection Program

Appendix C. Evidence Tables

Evidence Table 1A. Screening Among Older Women (cont'd)

Participants	Screening Program	Outcomes & Measures	Relevant Results	Quality Considerations
N = not defined (only rates given)	National screening program with call and recall of ages 25-69 (1970-1987) now 20-69 (began 1988) for screening q2-3 years	Primary outcome = histologically proven lesions Secondary outcome = moderate to high grade smears	INTERVAL: Cumulative rate of histologically verified lesions among women referred for colposcopy rises from 10% of those referred within 2 yrs of a normal Pap to 70% of those referred within 5 yrs. AGE: After age 60, the rate of moderate to high-grade dysplasia among women without prior screening = 16/1000; 5.7/1000 if 1-4 prior screens; 2.8 if ≥ 5	Denom=2 Attrition=1 Indications=1 Interval=2 Age=2 Relevance=1 Standard=2 Score=11 Grade=Fair
N = 128,805 women with initial normal Pap and another within 36 months (excludes those with glandular cell abnormalities) Age <30 9.5% 30-49 41.6% 50-64 36.8% ≥ 65 12.1% White 57.2% Black, non-Hisp 13.4% Hisp 17.6% Asian 2.4% AmInd 8.5%		Primary outcome = high grade SIL or worse on cytology Secondary outcome = other cytologic abnormality Second smears results: (%) Benign 94.4 ASCUS 3.4 LSIL 0.9 HSIL 0.2 Cancer 0.0	AGE: 2nd screening results LSIL (%) HSIL (%) SCCA (%) < 30 389 (3.2) 80 (0.7) 1 (0.01) 30-49 523 (1.0) 111 (0.2) 5 (0.01) 50-64 186 (0.4) 67 (0.14) 4 (0.01) ≥ 65 42 (0.3) 13 (0.08) 3 (0.02) INTERVAL: Age-adjusted incidence per 10,000 women Time from normal ASCUS LSIL HSIL + 9-12 mos 377 107 25 13-24 373 125 29 25-36 415 141 33 P (trend) 0.36 0.01 0.42	Denom=2 Attrition=2 Indications=1 Interval=2 Age=2 Relevance=2 Standard=0 Score=11 Grade=Fair

Appendix C. Evidence Tables

Evidence Table 1A. Screening Among Older Women (cont'd)

Source: Author, Year	Study Design & Characteristics	Stated Objective(s)	Location & Time Period	Data Source(s)
Sawaya et al., 2000[36]	Prospective cohort	To determine differences in incidence of high-grade cytology following 1, 2, or 3 or more normal smears	US 1991-1999	Prospective data collection as part of National Breast & Cervical Cancer Early Detection Program

Appendix C. Evidence Tables

Evidence Table 1A. Screening Among Older Women (cont'd)

Participants	Screening Program	Outcomes & Measures	Relevant Results			Quality Considerations
N = 128,805 women with initial normal Pap and at least another in the timeframe Mean age = 49.9 13.3 yrs		Primary outcome = HSIL+ on cytology	AGE: Incidence of abnormalities within 36 month following 1, 2, or 3 consecutive normal smears			Denom=2 Attrition=2 Indications=1 Interval=2 Age=2 Relevance=2 Standard=0 Score=11 Grade=Fair
				LSIL %	HSIL+ %	

			<30 (1)
			         (2)
			         (3)
				3.47
				2.33
				1.70
					0.65
					0.67
					0.62	

			30-39	1.48 0.90 0.40	0.44 0.11 0.11	
			40-49	0.85 0.75 0.40	0.18 0.16 0.02	
			50-59	0.49 0.31 0.22	0.19 0.10 0.06	
			≥ 60	0.35 0.30 0.14	0.12 0.09 0.02	

			PRIOR HISTORY:
			Age & interval adjusted relative risk
			#prior normals RR/1,000 (95%CI)
			                              women
			       1 23.6
			       2 16.5 (12.0-22.4)
			       3+ 7.6 (4.0-14.2)			

Appendix C. Evidence Tables

Evidence Table 1A. Screening Among Older Women (cont'd)

Source: Author, Year	Study Design & Characteristics	Stated Objective(s)	Location & Time Period	Data Source(s)
Sawaya et al., 2000[44]	Prospective cohort	To determine the positive predictive value of cervical smears in previously screened post-menopausal women and to determine the effect of oral estrogen plus progestin on incident cervical cytologic abnormalities	US study 20 clinics Date Not Specified	Data collected prospectively for the Heart & Estrogen-progestin Replacement Study (HERS) a randomized, placebo controlled trial of hormone replacement

Appendix C. Evidence Tables

Evidence Table 1A. Screening Among Older Women (cont'd)

Participants	Screening Program	Outcomes & Measures	Relevant Results	Quality Considerations

N = 2,561 women with normal baseline smears
 	
Mean age: 66
Nonwhite: ~ 10%
	Annual screening during an average of 4 years of study participation	Primary outcome = "new cytologic abnormality" Defined as ASCUS, AGCUS, LSIL, or HSIL
		
		OUTCOME RATE
		
		Secondary outcome Results of evaluation for abnormal Pap
			Timing of abnormal smears after prior normal
			1 year later 78/2561 (3.0%)
			2 years later 32/2346 (1.4%)
			Total 110 abnormals
			
			                        Final Diagnosis
			            NL LSIL HSIL Endo* UKN
			Cytology
			ASCUS 66 2 1 1 4
			AGCUS 20 2 --- --- 1
			LSIL 8 2 --- --- 2
			HSIL --- --- 1 --- ---
			Total 94 6 2 1 7
			* Endometrial hyperplasia without atypia
				Denom=2
				Attrition=2
				Indications=2
				Interval=1
				Age=0
				Relevance=2
				Standard=1
				Score=10
				
				Grade=Fair

Appendix C. Evidence Tables

Evidence Table 1B. Pap Testing After Hysterectomy

Source: Author, Year	Study Design	Stated Objective(s)	Location & Time Period	Data Source(s)
Fox et al., 1999[47]	1. Population-based cross sectional study 2. Nested case-control study	To determine the risk of cytologic abnormality on a screening Pap for women >= 50 years with and without a uterine cervix	Dane County, WI 1/1/95-8/15/95	Regional cytology center database; physician-reported hysterectomy status
Pearce, et al., 1996[48]	Retrospective cohort study	To determine the prevalence and clinical importance of abnormal vaginal Pap smears in a population of women with hysterectomy	New Orleans, LA, 1/1/92-12/31/94	Charity Hospital cervical screening database; review of medical records for diagnostic test results following abnormal Paps

Appendix C. Evidence Tables

Evidence Table 1B. Pap Testing After Hysterectomy

Participants	Screening Program	Outcomes & Measures	Relevant Results	Quality Considerations
N=21,152 Pap smears (cross sectional study) 5330 s/p hysterectomy N=173 nested case-control study Age >= 50 years; women being followed up for previous abnormality or with bleeding/discharge excluded Population base is 97% white	Not stated hysterectomy	Primary outcome = abnormal Pap, defined as ASCUS, dysplasia (low and high grade SIL) or carcinoma as determined by cytopathologist (n=173) and reported by hysterectomy status For case-control study, primary exposure= hysterectomy status	Results from cross-sectional study: Rates of abnormality among women Among all 8.2/1000 Among those with hysterectomy 1.7/1000 Among those without hysterectomy 10.4/1000 From case-control study: Odds ratios for an abnormal screening Pap=0.09 (95% CI=0.02-0.24) for those with hysterectomy versus those without	Grade=Fair
N=9610 smears among women with hysterectomy and without history of cervical, vaginal, ovarian, fallopian tube or uterine cancer, or CIN III Mean age=52 83% black race	Not stated	Primary outcome= abnormal Pap, defined as ASCUS, LGSIL, HGSIL, or squamous-cell carcinoma; classified by two independent cytopathologists following review 27/79 women with abnormal results verified with colposcopy	27/79 women with abnormal Paps referred for colposcopy, 5/27 found to have biopsy-proved VAIN on colposcopy (PPV of Pap for VAIN detection 6.3%, 95% CI 3.1 to 18.0); no biopsy proved cases of carcinoma	Grade=Fair

Appendix C. Evidence Tables

Evidence Table 2. New Methods for Preparing or Evaluating Cervical Cytology

Source: Author, Year	Study Design & Characteristics	Interventions	Location & Time Period
AutoPap			
Colgan et al., 1995[83]	Prospective blinded comparison	AutoPap 300 QC rescreening vs. manual rescreening among smears initially categorized as negative	Ontario, Canada Time period not specified Independent service cytopathology laboratory
Patten et al., 1997b[84]	Diagnostic test evaluation No reference standard. No histological verification.	AutoPap 300 QC vs. initial reading	US 1997 Commercial or hospital/academic laboratories
Lee et al., 1998[85]	Diagnostic test evaluation	AutoPap primary screening system	9 cytopathology laboratories c. 1997
Stevens et al., 1997[86]	Prospective comparison	AutoPap 300 QC vs. random manual rescreening	Australia 1995-1995 Cytology laboratory

Appendix C. Evidence Tables

Evidence Table 2. New Methods for Preparing or Evaluating Cervical Cytology (cont'd)

Patients & Methods	Outcomes Measured	Study Results & Limitations*	Quality Considerations
AutoPap			
3487 smears screened manually as negative	Blind interpretation No. of FN/No. rescreened No. of FN with AutoPap/total FN GS: cytology (independent panel) applied to discrepant cases	AutoPap 20% review rate Yield: (ASCUS+) = 57/3487 (1.6%) (LSIL+) = 10/3487 (0.29%) FN (AutoPap)/FN total (ASCUS+) = 57/86 (66.3%) FN (AutoPap)/total (LSIL+) = 10/13 (76.9%)	Quality Score=6 Ref. Std: 0 Blind: 2 Verification: 1 Consecutive: 1 Spectrum: 1 Publication: 1 Industry: 0
Current archive sensitivity: 2339 Pap smears including positives and negative controls; Historic sensitivity study 3028 Pap smears comprising positive and negative controls	No. of FN detected by AutoPap/total FN (estimate of sensitivity)	AutoPap 10% review rate: Est. Se (ASCUS+) = 67/203 (33%) Est. Se (LSIL+) = 18/32 (56%) AutoPap 20% review rate: Est. Se (ASCUS+) = 103/203 (51%) Est. Se (LSIL+) = 21/32 (66%) No data on specificity of AutoPap rescreen	Quality Score=4 Ref. Std: 0 Blind: 2 Verification: 0 Consecutive: 0 Spectrum: 1 Publication: 1 Industry: 0
683 Pap smears (264 normal)	Sensitivity and workload reduction GS: cytology (independent panel)	Se (ASCUS+) = 330/357 (92.4%) Se (LSIL+) = 195/199 (98%) Feasibility study of location-guided screening. No results for slides initially read as normal.	Quality Score=6 Ref. Std: 0 Blind: 2 Verification: 2 Consecutive: 0 Spectrum: 1 Publication: 1 Industry: 0
1840 Pap smears initially read as "normal"	No. of FN detected by AutoPap/total FN (estimated sensitivity) GS: cytology (independent panel review of discrepant cases)	AutoPap: Est. Se: 30% review rate: 4/7 (52%) 20% review rate 3/7 (43%) 10% manual review: Est. Se 0/7 = 0%	Quality Score=6 Ref. Std: 0 Blind: 2 Verification: 1 Consecutive: 1 Spectrum: 1 Publication: 1 Industry: 0

Appendix C. Evidence Tables

Evidence Table 2. New Methods for Preparing or Evaluating Cervical Cytology (cont'd)

Source: Author, Year	Study Design & Characteristics	Interventions	Location & Time Period
AutoPap			
Wilbur et al., 1996[87]	Diagnostic test evaluation	AutoPap 300 QC 10% or 20% rescreening rate	New York Commercial and academic cytopathology laboratories
Wilbur et al., 1998[88]	Prospective two-armed comparison	AutoPap primary screening system vs. manual screening	Time period not specified 5 commercial cytopathology laboratories
Wilbur et al., 1999[89]	Prospective two-armed comparison of conventional manual cytologic review with 10% rescreening	Rescreen of conventional manual screening of Pap using AutoPap	Five commercial laboratory sites Time period not given

Appendix C. Evidence Tables

Evidence Table 2. New Methods for Preparing or Evaluating Cervical Cytology (cont'd)

Patients & Methods	Outcomes Measured	Study Results & Limitations*	Quality Considerations†
AutoPap			
86 cases of known HSIL+ from archives of two labs	Sensitivity of AutoPap to detect known HSIL+ GS: histology	In 86 biopsy–positive cases with cytological diagnosis of HSIL+, AutoPap selected 66 (77%) at 10% review fraction and 74 (86%) at 20% review fraction Limitation: not rescreening use	Quality Score=4 Ref. Std: 2 Blind: 0 Verification: 1 Consecutive: 0 Spectrum: 0 Publication: 1 Industry: 0
25,124 Pap smear slides, excluding "high-risk" slides	Estimated sensitivity Relative TPR Relative FPR GS: cytology (independent panel review of discrepant cases). No histological verification.	Est Se (ASCUS+) = 1199/1397 (85.8%) Est Se (LSIL+) = 321/348 (92.2%) Relative TPR (AutoPap/conv) = 1199/1106 = 1.08 Relative FPR (AutoPap/conv) = 1123/1322 = 0.85 Limitation: results not reported for performance of AutoPap System on slides with manual screen witin normal limits. AutoPap system uses a different algorithm from AutoPap 300 QC	Quality Score=6 Ref. Std: 0 Blind: 2 Verification: 1 Consecutive: 1 Spectrum: 1 Publication: 1 Industry: 0
25,124 slides, excluding "high risk" slides	Pap results: Bethesda method Reference standard: HGSIL and above	Prevalence: HSIL+ = 70/25,124=0.28% AutoPap: Se (ASCUS+, HSIL+)=86% Se (LSIL+, HSIL+)=92% Se (HSIL+, HSIL+)=97% Manual reading: Se (ASCUS+, HSIL+)=79% Se (LSIL+, HSIL+)=86% Se (HSIL+, HSIL+)=93% Specificity cannot be calculated	Quality Score= 5 Ref. Std:0 Blind:2 Verification:0 Consecutive:0 Spectrum:1 Publication:1 Industry:1

*Format for display of results of conventional Pap test: Prevalence (GS threshold)=disease/total=%; Sensitivity (Pap threshold/GS threshold)=true positive/disease=%; Specificity (Pap threshold/GS threshold)=true negatives/non-diseased=%

Appendix C. Evidence Tables

Evidence Table 2. New Methods for Preparing or Evaluating Cervical Cytology (cont'd)

Source: Author, Year	Study Design & Characteristics	Interventions	Location & Time Period
Papnet			
Ashfaq et al., 1995[90]	Diagnostic test evaluation Prospective validation of Papnet rescreening	Papnet rescreening vs. manual rescreening among samples initially categorized as negative	Dallas, TX Time period not specified Academic medical center
Slagel et al., 1995[91]	Diagnostic test evaluation	Papnet screening	Iowa City, IA 1990
Farnsworth et al., 1996[92]	Prospective, blind comparison	Papnet rescreening vs. manual rescreening	Sydney, Australia 1/95-9/95 Large cytopathology laboratory

Appendix C. Evidence Tables

Evidence Table 2. New Methods for Preparing or Evaluating Cervical Cytology (cont'd)

Patients & Methods	Outcomes Measured	Study Results & Limitations*	Quality Considerations
Papnet			
2238 smears screened manually as negative. No patients with known prior dysplasia, radiation or chemo-therapy, post-menopausal vaginal bleeding; symptomatic	No. of FN/ No rescreened GS: cytology (single independent cytopathologist) only for slides called unsatisfactory or atypical on Papnet review	Papnet rescreening resulted in 91/2238 "review," 45/2238 "inadequate," 2102/2238 negative (94%) Manual rescreen of 91 atypical cases found: 5/91 ASCUS or LSIL. 86/91 negative (20% BCC) Detection rate 0.2% Manual rescreen of 45 inadequate found: 36/45 negative, 9 truly unsatisfactory	Quality Score=5 Ref. Std: 0 Blind: 0 Verification: 1 Consecutive: 1 Spectrum: 1 Publication: 1 Industry: 1
500 previously screened Pap smears; 435 of which had been screened previously as normal	No. of FN/No. rescreened GS: cytology (single independent pathologist). Discrepant cases reviewed	FN yield: 15/450 (3.3%)	Quality Score=7 Ref. Std: 0 Blind: 2 Verification: 1 Consecutive: 1 Spectrum: 1 Publication: 1 Industry: 1
54,658 Pap smears initially read as normal, and 1022 Pap smears classified as abnormal	No. of FN/No. rescreened GS: cytology (single cytopathologist). Histological validation for small subset of Papnet positive smears	Papnet identified 266/54658 = 0.49% (32 ASCUS, 217 LSIL, 17 HSIL) Papnet detected all 122 HSIL and ICC, but failed to detect 112 (14 ASCUS, 98 LSIL) of 1022 known positive smears	Quality Score=7 Ref. Std: 0 Blind: 2 Verification: 1 Consecutive: 1 Spectrum: 1 Publication: 1 Industry: 1

Appendix C. Evidence Tables

Evidence Table 2. New Methods for Preparing or Evaluating Cervical Cytology (cont'd)

Source: Author, Year	Study Design & Characteristics	Interventions	Location & Time Period
Papnet			
Ashfaq et al., 1997[93]	Diagnostic test evaluation Prospective, blind comparison of Papnet	Papnet (pre-) screening vs. manual screening diagnosis	Dallas, TX Time period not specified Academic medical center
Duggan and Brasher, 1997[94]	Prospective, blind comparison of Papnet vs. manual screening	Papnet pre-screening vs. manual screening	Alberta, Canada 3-month period in 1995 Pathology laboratory
Halford et al., 1997[95]	Prospective comparison of Papnet rescreening vs. historical rapid rescreening No independent cytology panel. No histological verification	Papnet rescreening vs. rapid manual rescreening	Australia Time period not specified

Appendix C. Evidence Tables

Evidence Table 2. New Methods for Preparing or Evaluating Cervical Cytology (cont'd)

Patients & Methods	Outcomes Measured	Study Results & Limitations*	Quality Considerations
Papnet			
5170 consecutive Pap smears	Sensitivity and specificity to known abnormals GS: cytology (single independent cytopathologist) for discrepant cases only	Papnet: Se 86.1% Sp (ASCUS+) = 82% Sp (LSIL+) = 93.6% Manual screening: Se = 77.3% Sp (ASCUS+) = 77.3% Sp (LSIL+) = 92.5% Performance among manually screened negative slides not reported	Quality Score=7 Ref. Std: 0 Blind: 2 Verification: 1 Consecutive: 1 Spectrum: 1 Publication: 1 Industry: 1
5037 consecutive Pap tests for cervical cancer screening (smears from colposcopy or gynecology-oncology clinics excluded), 4574 of which were negative on manual screening	No. of FN/No. rescreened GS: cytology (panel). Verification of discrepant cases	Papnet: 17/4574 = 0.37%	Quality Score=7 Ref. Std: 0 Blind: 2 Verification: 1 Consecutive: 1 Spectrum: 1 Publication: 1 Industry: 1
1000 negative smears seeded with 20 "difficult" cases	No. of FN/No. rescreened Sensitivity to known abnormals	Results not given for false negative detection Se(known abn ASCUS+) 19/20 (95%) Rapid manual rescreening Se (known abn ASCUS+) = 9/20 (45%)	Quality Score=2 Ref. Std: 0 Blind: 0 Verification: 0 Consecutive: 0 Spectrum: 0 Publication: 1 Industry: 1

Appendix C. Evidence Tables

Evidence Table 2. New Methods for Preparing or Evaluating Cervical Cytology (cont'd)

Source: Author, Year	Study Design & Characteristics	Interventions	Location & Time Period
Papnet			
Jenny et al., 1997[96]	Diagnostic test evaluation	Papnet	Zurich, Switzerland 1988-1994 Private cytopathology laboratory
Kaufman et al., 1998[97]	Diagnostic test evaluation, Papnet rescreening compared with colposcope-directed biopsy All subjects verified	Rescreening of conventional Pap smear using Papnet	Texas 1993-1995
Mango and Valente, 1998[98]	Diagnostic test evaluation	Papnet rescreening vs. manual rescreening	1985-1992 Academic cytopathology laboratories

Appendix C. Evidence Tables

Evidence Table 2. New Methods for Preparing or Evaluating Cervical Cytology (cont'd)

Patients & Methods	Outcomes Measured	Study Results & Limitations*	Quality Considerations
Papnet			
1200 smears 516 with histologically proven diagnosis; 29 of which had been manually screened as negative	No. of FN/No. rescreened GS: Histological validation. No validation of test negatives	Papnet rescreening: Se(LSIL+) = 26/29 = 89% In primary screening: 2 independent Papnet reviews identified 435 & 469 of 516 (91% & 84%) Manual screening: 403/561 (78.1%) Relative TPR 1.08-1.16 Papnet compared to manual screening	Quality Score=7 Ref. Std: 2 Blind: 2 Verification: 1 Consecutive: 0 Spectrum: 0 Publication: 1 Industry: 1
160 women with colposcopy and biopsy within 1 year of a smear initially reported as ASCUS	Blind interpretation Papnet diagnosis (ASCUS/ AGUS, LSIL, HSIL, Ca) GS: colposcope-directed biopsy (normal, HPV, CIN 1, CIN 2, CIN 3)	Prevalence(CIN1)=69/160=43% Se (LSIL,CIN1)=26/69=38% Sp (LSIL,CIN1)=84/91=92% Prevalence(CIN2/3)=22/160= 14% Se (LSIL,CIN2/3)=9/22=41% Sp (LSIL,CIN2/3)=114/138=83% Limitations: Narrow spectrum of disease (ASCUS/AGUS)	Quality Score=10 Ref. Std: 2 Blind: 2 Verification: 2 Consecutive: 2 Spectrum: 1 Publication: 1 Industry: 0
2293 smears underwent Papnet rescreening; 13761 underwent manual rescreening	No. of FN/No. rescreened GS: cytology (single pathologist). No verification of rescreen negatives. No histological validation	Papnet: FN yield (ASCUS+)142/2293 (6.3%) FN yield (LSIL+) 48/2293 (2.1%) Manual rescreen: FN yield (ASCUS+) 82/13761 (0.6%) FN yield (LSIL+) 36/13761 (0.3%)	Quality Score=4 Ref. Std: 0 Blind: 0 Verification: 1 Consecutive: 1 Spectrum: 1 Publication: 1 Industry: 0

Appendix C. Evidence Tables

Evidence Table 2. New Methods for Preparing or Evaluating Cervical Cytology (cont'd)

Source: Author, Year	Study Design & Characteristics	Interventions	Location & Time Period
Papnet			
Mitchell and Medley, 1998[99]	Evaluation of Papnet rescreening	Papnet rescreening	Australia 1990-1992
O'Leary et al., 1998[100]	Prospective evaluation of Papnet rescreening	Papnet rescreening	Washington, DC 1994-1995 Government pathology organization
Duggan, 2000[60]	Diagnostic test evaluation of Papnet system vs. conventional manually read Paps in primary screening Reference standard: consensus by peer review	Rescreen of conventional manual screening Pap results using Papnet	Calgary, Alberta Starting date for slide selection: December 1995

Appendix C. Evidence Tables

Evidence Table 2. New Methods for Preparing or Evaluating Cervical Cytology (cont'd)

Patients & Methods	Outcomes Measured	Study Results & Limitations*	Quality Considerations
Papnet			
19,805 Pap smears with 2 negative manual screens	No. of FN/No. rescreened GS: cytology (panel). No verification of rescreen negatives. Histologic validation of some rescreen positives	FN yield (single cytologist): 212/19805 (1.1%) FN yield (cytology panel): 162/212 (0.82%) FN yield (histology): 14/26 high-grade and 33/102 low-grade abnormalities histologically confirmed	Quality Score=5 Ref. Std: 0 Blind: 0 Verification: 1 Consecutive: 1 Spectrum: 1 Publication: 1 Industry: 1
5478 consecutive Pap smears read as normal or benign cell changes on initial and random rescreening	No. FN/No. rescreened GS: cytology (independent panel). Verification of discrepant cases only	Papnet identified 1614 (29%) of slides requiring review; 1166 because no endocervical component was identified. 448 (8% of total) Only 11 cases demonstrated abnormal cells (5 ASCUS, 1 AGUS, 6 LSIL+) 11/5478 (0.2%)	Quality Score=5 Ref. Std: 0 Blind: 0 Verification: 1 Consecutive: 1 Spectrum: 1 Publication: 1 Industry: 1
2,200 archival slides selected: 2000 normals, 200 abnormals	All Pap results: Bethesda method Reference standard: Negative, ASCUS/AGUS, LSIL, HSIL, Ca	Prevalence (artificially set): LSIL=52/2195=2.4% HSIL=62/2195=2.8% Ca=39/2195=1.8% Se (ASCUS/AGUS, LSIL+)=72.7% Sp (ASCUS/AGUS, LSIL+)=99.4% Se (LSIL, LSIL+)=87.2% Sp (LSIL, LSIL+)=99.9% Se (HSIL, HSIL+)=98.2% Sp (HSIL, HSIL+)=99.9%	Quality Score= 7 Ref. Std:1 Blind:2 Verification:1 Consecutive:0 Spectrum:1 Publication:1 Industry:1

Appendix C. Evidence Tables

Evidence Table 2. New Methods for Preparing or Evaluating Cervical Cytology (cont'd)

Source: Author, Year	Study Design & Characteristics	Interventions	Location & Time Period
Papnet			
Kok et al., 2000[101]	Diagnostic test evaluation of neural network-based screening using Papnet compared to traditional Pap Reference standard: 69 patients with biopsy confirmed carcinoma	Women randomized to receive screening by Papnet (245,527) or traditional Pap (109,104) Collection method: Cytobrush	Netherlands 1992-1997
ThinPrep 2000			
Corkill et al., 1997[102]	Diagnostic test evaluation Prospective, split sample, double-masked trial. Discrepancies and 5% of test-negatives verified by single independent pathologist only for second group. No histologic verification of test-positives	Conventional Pap vs. ThinPrep. Split sample collected with cytobrush/spatula	Colorado 5/95-6/95 and 3/96-4/96 11 planned parenthood clinics
Lee et al., 1997[103]	Diagnostic test evaluation, conventional Pap vs. ThinPrep ThinPrep clinical trial: 6 centers, prospective, split sample, double-masked trial Discrepancies and 5% of test-negatives verified by single independent pathologist No histologic verification of test-positives	Conventional Pap vs. ThinPrep. Split sample collected with broom device	US 1996 6 sites; 3 community clinics and 3 hospitals

Appendix C. Evidence Tables

Evidence Table 2. New Methods for Preparing or Evaluating Cervical Cytology (cont'd)

Patients & Methods	Outcomes Measured	Study Results & Limitations*	Quality Considerations
Papnet			
Women undergoing routine screening aged 30 to 60 years (194,358), or screening for any other reason (160,373).	All Pap results: abnormal defined as CIN-I or greater Reference standard: biopsy-confirmed squamous cell carcinoma	Prevalence SCC=71/354,631=0.02% Estimated Se (CIN-I by Papnet+, SCC)=90.4% Specificity cannot be calculated No gold standard verification of all test positives and test negatives; specificity cannot be calculated	Quality Score=9 Ref. Std: 2 Blind: 2 Verification: 0 Consecutive: 2 Spectrum: 1 Publication: 1 Industry: 1
ThinPrep 2000			
462 subjects in first group, 1239 in following group. All patients >18 years, premenopausal, and no history of abnormal Pap	Concordance between ThinPrep and conventional smear; discrepancies and 5% of test-negatives verified by single independent pathologist only for second group	Unable to get estimate of specificity because no test positives verified with histology. ThinPrep more likely to identify slides as abnormal (LSIL+) than conventional smear	Quality Score=3.5 Ref. Std: 0 Blind: 2 Verification: 0 Consecutive: 0 Spectrum: 1 Publication: 1 Industry: .5
6747 women, recruited from 6 centers. 3 hospital centers selected for high proportion of high-risk women. 58.9% white, 21.7% black, 13.9 % Hispanic, 2.2% other	Concordance between ThinPrep and conventional smear; discrepancies and 5% of test-negatives verified by single independent pathologist	Unable to get estimate of specificity because no test positives verified with histology. ThinPrep more likely to identify slides as abnormal (ASCUS or LSIL) than conventional smear in screening centers, but not in hospitals	Quality Score=5.5 Ref. Std: 0 Blind: 2 Verification: 1 Consecutive: 0 Spectrum: 1 Publication: 1 Industry: .5

Appendix C. Evidence Tables

Evidence Table 2. New Methods for Preparing or Evaluating Cervical Cytology (cont'd)

Source: Author, Year	Study Design & Characteristics	Interventions	Location & Time Period
ThinPrep 2000			
Roberts et al, 1997[104]	Diagnostic test evaluation. split sample, ThinPrep vs. conventional Pap Discrepant results reviewed Most of test-positives with HSIL verified. No test negatives verified with histology	Split sample technique with Cervex broom used to prepare both conventional Pap and ThinPrep 2000	Australia 1996–1997
Bolick and Hellman, 1998[105]	Diagnostic test evaluation Selection of negative smears uncertain	Conventional Pap (broom-type sampling device or a combination endocervical brush and plastic spatula) vs. ThinPrep	Utah 1996-1997

Appendix C. Evidence Tables

Evidence Table 2. New Methods for Preparing or Evaluating Cervical Cytology (cont'd)

Patients & Methods	Outcomes Measured	Study Results & Limitations*	Quality Considerations
ThinPrep 2000			
35,560 pairs received from 500 practitioners who chose to offer ThinPrep along with conventional smear. 24% of women undergoing screening pap, and 35% of women needing referral pap had both samples taken	No blind interpretation Conventional smear and ThinPrep thresholds of minor changes/HPV, LSIL, inconclusive abnormality, HSIL, cytologic reference standard for test negatives. Histologic reference standard for those with HSIL or inconclusive results on either conventional smear or ThinPrep	All discrepant results reviewed by independent cytopathologist Most of HSIL and inconclusive results on either ThinPrep or conventional smear had histologic confirmation No test negatives verified, or those with LSIL Relative True Positive rate (ThinPrep)=178/158=1.13 Relative False Positive rate(ThinPrep)=37/33=1.12	Quality Score=1.5 Ref. Std: 0 Blind: 0 Verification: 0 Consecutive: 0 Spectrum: 0 Publication: 1 Industry: .5
39,408 conventional Paps, and 10,694 ThinPrep specimens, collected concurrently in two patient populations, described only by age	No blind interpretation Pap and ThinPrep specimen with threshold of LSIL GS: histology with threshold of "positive"	Screening test: ThinPrep Prevalence(not specified)= 42/54=78% Se (LSIL,NS)=40/42=95% Sp (LSIL,NS)=7/12=58% Screening test: Pap Prevalence(NS)=67/89=75% Se (LSIL,NS)=57/67=85% Sp (LSIL,NS)=8/22=36%	Quality Score=4 Ref. Std: 2 Blind: 0 Verification: 0 Consecutive: 0 Spectrum: 0 Publication: 1 Industry: 1

Appendix C. Evidence Tables

Evidence Table 2. New Methods for Preparing or Evaluating Cervical Cytology (cont'd)

Source: Author, Year	Study Design & Characteristics	Interventions	Location & Time Period
ThinPrep 2000			
Inhorn et al., 1998[106]	Diagnostic test evaluation of ThinPrep vs. conventional Pap Cancer specimens from ThinPrep clinical trial, including results from ThinPrep beta and ThinPrep 2000 ThinPrep and conventional smear obtained from known cancer patients, split-sample No disease negatives studied	ThinPrep vs. conventional Pap. Split-sample, obtained with broom device	5 institutions, gynecologic oncology clinics
Papillo et al., 1998[107]	Diagnostic test evaluation ThinPrep slides compared with historical cohort of conventional Pap slides for specimen adequacy and yield of abnormal diagnoses Histologic verification of most ThinPrep HSIL. No verification of test-negatives	ThinPrep vs. conventional smear ThinPrep Direct to vial	Vermont 1997 12 practice sites that chose to offer ThinPrep

Appendix C. Evidence Tables

Evidence Table 2. New Methods for Preparing or Evaluating Cervical Cytology (cont'd)

Patients & Methods	Outcomes Measured	Study Results & Limitations*	Quality Considerations
ThinPrep 2000			
47 patients with known cervical cancer	No blind interpretation Ability of ThinPrep and conventional smear to identify cancerous cells as cancer	Independent reference pathologist only for samples from ThinPrep 2000 study only Study only evaluated cancer patients, therefore unable to estimate specificity ThinPrep identified 21/22 smears as cancer (1 was ASCUS) Conventional smear identified 19/22 as cancer (1 was ASCUS, 2 were unsatisfactory)	Quality Score=3.5 Ref. Std: 2 Blind: 0 Verification: 0 Consecutive: 0 Spectrum: 0 Publication: 1 Industry: .5
Uncertain sample collection. Spectrum not described	Specimen adequacy and yield of abnormal diagnoses Thresholds for ThinPrep and conventional smear of ASCUS/AGUS, LSIL, SIL NOS, HSIL	No reference standard; compared only to historical cohort of conventional smear samples. Prevalence of disease unknown Histologic verification of most ThinPrep HSIL ThinPrep resulted in more abnormal diagnoses (ASCUS+) than historical conventional smear, and less unsatisfactory or limited smears PPV of ThinPrep for HSIL=93.2% Historical Positive Predictive Value of conventional smear for HSIL=78.8%	Quality Score=1.5 Ref. Std: 0 Blind: 0 Verification: 0 Consecutive: 0 Spectrum: 0 Publication: 1 Industry: .5

Appendix C. Evidence Tables

Evidence Table 2. New Methods for Preparing or Evaluating Cervical Cytology (cont'd)

Source: Author, Year	Study Design & Characteristics	Interventions	Location & Time Period
ThinPrep 2000			
Sherman et al., 1998[108]	Diagnostic test evaluation, conventional Pap vs. ThinPrep Data from ThinPrep clinical trial All abnormals, and random 5% of test-negatives verified by independent cytologists and pathologists. No histologic verification of test-positives. Diagnoses compared with HPV results	Conventional Pap vs. ThinPrep. Split sample collected with broom device	US 1996 6 sites; 3 community clinics and 3 hospitals

Appendix C. Evidence Tables

Evidence Table 2. New Methods for Preparing or Evaluating Cervical Cytology (cont'd)

Patients & Methods	Outcomes Measured	Study Results & Limitations*	Quality Considerations
ThinPrep 2000			
1954 slides: 895 discrepant results, 759 concordant positives, and 300 random negatives from clinical trial	Blind interpretation Sensitivity of ThinPrep and conventional smear as compared to most abnormal reference cytology diagnosis PPV estimated by comparison with HPV type	Unable to get estimate of specificity because no test positives verified with histology Se ThinPrep (HSIL)=88.6% Se conventional smear (HSIL)=81.2% No difference in HPV detection in smears called abnormal between conventional smear and ThinPrep More cancer-associated HPV types seen in ThinPrep+ patients. Very few HPV+ in 300 random test-negatives	Quality Score=4.5 Ref. Std: 0 Blind: 2 Verification: 1 Consecutive: 0 Spectrum: 0 Publication: 1 Industry: .5

Appendix C. Evidence Tables

Evidence Table 2. New Methods for Preparing or Evaluating Cervical Cytology (cont'd)

Source: Author, Year	Study Design & Characteristics	Interventions	Location & Time Period
ThinPrep 2000			
Hutchinson et al, 1999[109]	Diagnostic test evaluation Verification of test negatives in random sample of 150 women Pap, ThinPrep® vs. colposcopy /biopsy (Other methods tested: Papnet®, cervioscopy but results not reported in this paper)	Conventional pap smear using split samples & ThinPrep® Collection method: Cervex Brush® (Unimar)	Guanacaste Costa Rica, time period not given
Weintraub and Morabia, 2000[49]	In community-based screening, diagnostic test evaluation of ThinPrep® compared to results obtained with standard Pap Reference standard: histology (among Pap +)	Results obtained by conventional Pap versus ThinPrep® ThinPrep® collection method: Cervex Brush® (Unimar)	Europe ThinPrep® samples obtained from 11/1/96-12/31/97; conventional 1/1/95-12/31/97

Appendix C. Evidence Tables

Evidence Table 2. New Methods for Preparing or Evaluating Cervical Cytology (cont'd)

Patients & Methods	Outcomes Measured	Study Results & Limitations*	Quality Considerations
ThinPrep 2000			
8636 in final analysis, non-virgin, non-pregnant women Aged (not given)	ThinPrep® results: ASCUS, LGSIL, HSIL, carcinoma GS: colposcopy-directed biopsy: reported as equivocal, LSIL, HSIL, cancer	Prevalence (LGSIL on colposcopy) =186/8636=2.2% Se (ASCUS ThinPrep®, LGSIL colp)=31.8% Sp ASCUS ThinPrep®, LGSIL colp)=93.2% Se (LGSIL ThinPrep®, LGSIL colp)=78.7% Sp (LGSIL ThinPrep®, LGSIL colp)=100% Prevalence (HGSIL on colposcopy) =126/8636=1.5% Se (LGSIL ThinPrep®, HGSIL colp)=71.9% Sp (LGSIL ThinPrep®, HGSIL colp)=100% Se (HGSIL ThinPrep®, HGSIL colp)=89.4% Sp (HGSIL ThinPrep®, HGSIL colp)=99.9% Limits: uncertain time between Pap and colposcopy	Quality Score=6 Ref. Std:1 Blind:0 Verification:1 Consecutive:2 Spectrum:1 Publication:1 Industry:1
39,864 ThinPrep® slides and 130,381 conventional Paps Reference standard results available for 509	Pap results: Bethesda Histologic diagnosis: negative, inconclusive, LSIL, HSIL	Prevalence: LSIL=117/509=23.0% HSIL=196/509=38.5% Ca=6/509=1.2% No gold standard verification of ThinPrep® negatives	Quality Score=8 Ref. Std:1 Blind:2 Verification:0 Consecutive:2 Spectrum:1 Publication:1 Industry:1

Appendix C. Evidence Tables

Evidence Table 3A. Performance of HPV Testing for Screening

Source: Author, Year	Study Design & Characteristics	Interventions	Location & Time Period
Cuzick et al., 1999[21]	Diagnostic test evaluation among older population undergoing routine screening for HPV and assessment of outcome by colposcopy Pap negatives not verified, HPV negatives verified except for Hybrid Capture where a sample of negative controls were evaluated	HPV testing of Pap results using consensus PCR/ SHARP detection for high risk subtypes, and Hybrid Capture I and Hybrid Capture II	40 gynecologic practices in UK Time period not specified
Petry et al., 1999[61]	Diagnostic test evaluation (HPV) among HIV+ population and assessment of outcome by colposcopy; study also contains an incidence arm	HPV testing of Pap results using Hybrid Capture I	Germany 1990-1998

Appendix C. Evidence Tables

Evidence Table 3A. Performance of HPV Testing for Screening (cont'd)

Patients & Methods	Outcomes Measured	Study Results & Limitations*	Quality Considerations
2,988 women 34 and older undergoing routine care	Cytology read as inadequate, negative, borderline, mild, moderate, severe, glandular atypia High risk HPV subtypes (16, 18, 31, 33, 35, 51, 522, 56, 58), other and negative Histology: Inadequate, Negative/no biopsy, borderline/CIN 1, CIN 2, CIN 3, Adeno in situ	Prevalence: Negative/no biopsy=2855/2988 =95.6% HPV/borderline=57/2988 =1.9% CIN 1=27/2988=0.9% CIN 2=8/2988=0.3% CIN 3=33/2988=1.1% Adeno in situ=1/2988=0.03% HC II: Se (High risk HPV, LGSIL+)=61.0% Sp (High risk HPV, LGSIL+)=95.1 % Se (High risk HPV, HGSIL+)=95.2 % Sp (High risk HPV, HGSIL+)=94.3 %	Quality Score=11 Ref. Std:2 Blind:2 Verification:2 Consecutive:2 Spectrum:1 Publication:1 Industry:1
138 HIV-infected women visiting gynecologic clinic for any reason	cytology: any atypia, >/= CIN I, >/= CIN II HPV: high risk types (16, 18, 31, 33, 35, 45, 51, 52, 56) vs. not histology: CIN II, CIN III or invasive cancer	Prevalence: CIN I=20/138=14.5% CIN II=5/138=3.6% CIN III=9/138=6.5% invasive ca=3/138=2.2% Se (High risk HPV, CIN II +)= 94.1% Sp (High risk HPV, CIN II +)= 70.3%	Quality Score=9 Ref. Std: 2 Blind: 0 Verification: 2 Consecutive: 2 Spectrum: 1 Publication: 1 Industry: 1

Appendix C. Evidence Tables

Evidence Table 3A. Performance of HPV Testing for Screening (cont'd)

Source: Author, Year	Study Design & Characteristics	Interventions	Location & Time Period
Schiffman et al., 2000[65]	Diagnostic test evaluation (HPV) among high incidence population and assessment of outcome by combining screening results with biopsy results Pap/ cervicography negatives verified by colposcopy in a 2% random sample	HPV testing of Pap results using Hybrid Capture Tube (HCT) and Hybrid Capture II (on a smaller subset of samples)	Guanacaste Province, Costa Rica 1993-1994
Womack et al, 2000a[63]	Diagnostic test evaluation (HPV) among population at high risk for HIV	HPV testing using Hybrid Capture II assay, probe B (13 high risk subtypes) and assessment of outcome by biopsy/ colposcopy	Chitungwiza and the greater Harare area, Zimbabwe c. 1999

Appendix C. Evidence Tables

Evidence Table 3A. Performance of HPV Testing for Screening (cont'd)

Patients & Methods	Outcomes Measured	Study Results & Limitations*	Quality Considerations
8554 women randomly selected in a door-to door survey	cytology: Bethesda system HPV: : high risk types (16, 18, 31, 33, 35, 45, 51, 52, 56 and 58) vs. not histology :negative, equivocal, low-grade lesion, high-grade lesion, cancer	Prevalence: Negative=7564/8554=88.4% Equivocal=661/8554=7.7% LG lesion=189/8554=2.2% HG lesion=128/8554=1.5% Cancer=12/8554=0.14% Se (High risk HPV by HCT, HG+)= 74.1% Sp (High risk HPV by HCT, HG+)= 93.4% Se (High risk HPV by HC II, HG+)= 88.4% Sp (High risk HPV by HC II, HG+)= 89.0% (more screening performance results available for different cutpoints of HPV positivity)	Quality Score=9 Ref. Std: 2 Blind: 0 Verification: 2 Consecutive: 2 Spectrum: 1 Publication: 1 Industry: 1
466 women aged 25 to 55 attending primary care clinics	HPV: high risk subtypes (16, 18, 31, 33, 35, 45, 51, 52, 56 and 58) vs. not Biopsy/colposcopy: Normal, low-grade lesion, high-grade lesion, cancer	Prevalence: Negative=350/466=75.1% LGSIL=60/466=12.9% HGSIL=56/466=12% for HIV+ women: Se (High risk HPV, LGSIL+)= 85.9% Sp (High risk HPV, LGSIL+)= 47.0% Se (High risk HPV, HGSIL)= 90.7% Sp (High risk HPV, HGSIL)= 41.3% for HIV- women: Se (High risk HPV, LGSIL+)=45.2% Sp (High risk HPV, LGSIL+)=75.3% Se (High risk HPV, HGSIL)= 61.5% Sp (High risk HPV, HGSIL)= 74.5%	Quality Score=9 Ref. Std: 2 Blind: 2 Verification: 2 Consecutive: 0 Spectrum: 1 Publication: 1 Industry: 1

Appendix C. Evidence Tables

Evidence Table 3A. Performance of HPV Testing for Screening (cont'd)

Source: Author, Year	Study Design & Characteristics	Interventions	Location & Time Period
Womack et al., 2000b[62]	Diagnostic test evaluation for primary screening	HPV testing using Hybrid Capture II assay, probe B (13 high risk subtypes) and assessment of outcome by biopsy/ colposcopy for all participants Testing of HPV viral load using relative light unit ratios Colposcopists blinded to results of HPV tests and visual inspection	Chitungwiza and the greater Harare area, Zimbabwe, Oct 1996-1997
Wright et al., 2000[64]	Diagnostic test evaluation (HPV) among population also screened using Pap, direct visual inspection, cervicography and assessment of outcome by biopsy/ colposcopy Abnormals on any test referred for verification	HPV testing using Hybrid Capture II assay in both patient collected and physician collected specimens	Environs of CapeTown, South Africa 1998-1999

Appendix C. Evidence Tables

Evidence Table 3A. Performance of HPV Testing for Screening (cont'd)

Patients & Methods	Outcomes Measured	Study Results & Limitations*	Quality Considerations
2140 women aged 25 to 55 attending primary care clinics	Primary outcome: High grade lesions or cancer (proved by biopsy and/or colposcopic impression) Low grade lesions (proved by biopsy and/or colposcopic impression) Secondary outcome: Light unit ratios of HPV positive specimens	Prevalence: LGSIL=561/2140=26.2% HGSIL=251/2140=10.1% Se (High risk HPV, LGSIL+) =64.0% Sp (High risk HPV, LGSIL+) =64.9% Se (High risk HPV, HGSIL+) = 80.9% Sp (High risk HPV, HGSIL+) = 61.6%	Quality Score=9 Ref. Std: 2 Blind: 2 Verification: 2 Consecutive: 0 Spectrum: 1 Publication: 1 Industry: 1
1365 previously unscreened black women aged 35 and older attending outpatient clinics	cytology: Bethesda system HPV: high risk subtypes (16, 18, 31, 33, 35, 45, 51, 52, 56 and 58) vs. not Colposcopy: biopsy-confirmed HGSIL, or cancer	Prevalence: LGSIL=40/1365=2.9% HGSIL=47/1365=3.4% Ca=9/1365=0.7% for clinican-collected specimens: Se (High risk HPV, LGSIL+)=81.3% Sp (High risk HPV, LGSIL+)=84.5% for self-collected specimens: Se (High risk HPV, LGSIL+)= 61.5% Sp (High risk HPV, LGSIL+)= 74.5% for clinican-collected specimens: Se (High risk HPV, HGSIL+)=81.3% Sp (High risk HPV, HGSIL+)=84.5% for self-collected specimens: Se (High risk HPV, HGSIL+)= 66.1% Sp (High risk HPV, HGSIL+)= 82.9%	Quality Score=9 Ref. Std: 2 Blind: 2 Verification: 2 Consecutive: 0 Spectrum: 1 Publication: 1 Industry: 1

Appendix C. Evidence Tables

Evidence Table 3B. Performance of HPV Testing for Triage

Source: Author, Year	Study Design & Characteristics	Interventions	Location & Time Period
Herrington et al., 1995[69]	Diagnostic test evaluation among abnormal Paps; screening for HPV compared with colpsocopy/ Histology Pap negatives not verified	HPV testing of abnormal Paps using two tests: Consensus PCR and in situ hybridization	Time period not specified
Sun et al., 1995[73]	Diagnostic test evaluation among women referred to a colposcopy clinic	HPV testing of abnormal Paps using Hybrid Capture I and consensus PCR	New York, NY or Montreal, Quebec 1992-1993

Appendix C. Evidence Tables

Evidence Table 3B. Performance of HPV Testing for Triage (cont'd)

Patients & Methods	Outcomes Measured	Study Results & Limitations*	Quality Considerations
167 patients with abnormal Paps referred for colposcopy	Primary outcome: Wart virus changes, CIN I, II, III or invasive carcinoma	Prevalence: WVC or CIN I =88/167=52.6% CIN II, III or ca =40/167=24.0% Se (PCR, ≥CIN II)=87.5% Sp (PCR, ≥CIN II)=62.2%	Quality Score=5 Ref. Std.:2 Blind: 0 Verification: 0 Consecutive:0 Spectrum:1 Publication:1 Industry:1
520 women referred to colposcopy clinic	Primary outcome: Low grade CIN High grade CIN or cancer Secondary outcome: Amount of HPV DNA by relative light unit reading	Prevalence: LG CIN = 161/520=31% HG CIN or ca= 105/520=20.2% Se (HC, ≥CIN I)=58.3% Sp (HC, ≥CIN I)=69.3% Se (HC, ≥CIN II)=66.7% Sp (HC, ≥CIN II)=60.7% Se (PCR, ≥CIN I)=84.6% Sp (PCR, ≥CIN I)=48.0% Se (PCR, ≥CIN II)=82.9% Sp (PCR, ≥CIN II)=34.9%	Quality Score=4.5 Ref. Std.:2 Blind: 0 Verification: 0 Consecutive:0 Spectrum:1 Publication:1 Industry:0.5

Appendix C. Evidence Tables

Evidence Table 3B. Performance of HPV Testing for Triage (cont'd)

Source: Author, Year	Study Design & Characteristics	Interventions	Location & Time Period
Bollen et al., 1997[66]	Diagnostic test evaluation among abnormal Paps; screening for HPV and type compared with colpsocopy/histology Pap negatives not verified, HPV negatives verified	HPV testing of abnormal Paps using two tests: CPI/IIG primer pair and MY09/11 polymerase chain reaction using SHARP signal system and probe sets A & B Colposcopy and histology	Amsterdam 1994-1995
Sigurdsson et al, 1997[67]	Descriptive study of HPV expression in 358 abnormal smears referred for colposcopy. Pap negatives not verified	1) Cytologic and histopathologic findings, 2) Presence and amount of HPV by hybrid capture and PCR, 3) Presence of HPV in swabs and biopsies, 4) Distribution of HPV type by cyto- and histopathological findings	Population based screening program, Iceland 1994

Appendix C. Evidence Tables

Evidence Table 3B. Performance of HPV Testing for Triage (cont'd)

Patients & Methods	Outcomes Measured	Study Results & Limitations*	Quality Considerations
190 consecutive mildly/moderately abnormal Paps referred for colposcopy	Blinded to HPV status while performing colposcopy HPV presence and type by CPI/IIG PCR, HPV presence by SHARP probe Histologic diagnosis: no dysplasia or LGSIL vs. HGSIL	Prevalence: LGSIL or none=134/190=70.5% HGSIL or none=56/190=29.5% Se (CPI/IIG, HSIL)=96% Sp (CPI/IIG)=33% Se (HPV subtypes 16,18,31,33,45/HGSIL)=68% Sp (HPV subtypes type16,18,31,33,45/HGSIL)=70% Se (SHARP Probe B, HSIL)=95% Sp (SHARP Probe B, HSIL)=40% Using both HPV tests: Se (SHARP Probe B +CPI/IIG, HSIL)=98% Sp (SHARP Probe B+CPI/IIG, HSIL)=28%	Quality Score=11 Ref. Std.:2 Blind:2 Verification:2 Consecutive:2 Spectrum:1 Publication:1 Industry:1
358 women diagnosed with and not treated earlier for abnormal smears referred for colposcopy	Cytology: Bethesda system Colposcope-directed biopsies and histology: negative, koilocytotic changes, CIN I, II, III, cancer HPV testing: Hybrid: high risk + intermediate risk=high risk (types 16, 18, 31, 33, 35, 45, 51, 52, and 56), vs. low risk (6, 11, 42, 43, 44) PCR: high risk (16, 18, 31, 33 and 35) vs. low (6, 11) or unclassified (other than these.)		Quality Score=8 Ref. Std.: 1 Blind: 2 Verification: 0 Consecutive: 2 Spectrum: 1 Publication: 1 Industry: 1

Appendix C. Evidence Tables

Evidence Table 3B. Performance of HPV Testing for Triage (cont'd)

Source: Author, Year	Study Design & Characteristics	Interventions	Location & Time Period
Adam et al., June 1998[71]	Diagnostic test evaluation among population w/2 minimally abnormal Paps or 1 highly abnormal; screening for HPV and type compared with colpsocopy/histology Pap negatives not verified, HPV negatives verified	HPV testing of abnormal Paps using PCR primer pair MY09/11 to identify high risk types 16, 18, 31, 33 and 35	Indigent population in Harris County, TX Time period not specified
Manos et al., May 1999[68]	Diagnostic test evaluation among population with ASCUS results from ThinPrep® Pap, screening for HPV and assessment of outcome by colposcopy Pap negatives not verified, HPV negatives verified	HPV testing of ASCUS ThinPrep® Pap results using Hybrid Capture II assay	Northern California HMO population 1995-1996
Hillemanns et al., 1999[72]	Diagnostic test evaluation (HPV) for detection of high grade lesions	Testing of self collected specimens and physician collected specimens for HPV using Hybrid Capture II	Munich, Germany Time period not specified

Appendix C. Evidence Tables

Evidence Table 3B. Performance of HPV Testing for Triage (cont'd)

Patients & Methods	Outcomes Measured	Study Results & Limitations*	Quality Considerations
1007 women with abnormal Paps referred for colposcopy	Pap diagnosis classified by Bethesda system HPV defined as negative consensus, 16, 18, 31&33&35, any high risk subtype, all multiple high risk subtypes, and unidentified type Histology: No CIN/ CIN 1/HPV, CIN 2, CIN 3/CIS, invasive carcinoma	Prevalence: No CIN=269/1007=27% CIN 1 w/HPV changes=477/1007=47% CIN II 124/1007=13% CIN III/CIS=12% Invasive carcinoma=4/1007=0.4% Se (High risk HPV, CIN II/III)=59% Sp (High risk HPV, CIN II/III)=59% Se (High risk HPV, CIN I +)=51% Sp (High risk HPV, CIN I +)=67%	Quality Score=9 Ref. Std.: 2 Blind: 2 Verification: 2 Consecutive: 0 Spectrum: 1 Publication: 1 Industry: 1
973 women undergoing routine screening with ASCUS Pap results and histologic diagnosis	ThinPrep® Pap diagnosis classified as ASCUS HPV + for high risk subtypes 16, 18, 31, 33, 35, 39, 45, 51, 52, 56, 58 Histology: normal, LSIL, HSIL, Cancer	Prevalence: Normal=783/973=80.4% LSIL=125/973=12.8% HSIL=64/973=6.7% Cancer=1/973=0.1% Se (High risk HPV, LGSIL+)=76.3% Sp (High risk HPV, LGSIL+)=69.5% Se (High risk HPV, HGSIL+)=89.2% Sp (High risk HPV, HGSIL+)=64.1%	Quality Score=11 Ref. Std.: 2 Blind: 2 Verification: 2 Consecutive: 2 Spectrum: 1 Publication: 1 Industry: 1
247 patients attending a colposcopy clinic	Primary outcome: CIN I, II, II, invasive cancer	Prevalence: CIN I =18/247=7.3% ≥CIN II=40/247=16.2% Physician collected samples: se (HPV, ≥CIN II+)=92.5% sp (HPV, ≥CIN II+)=72.5% Patient collected samples: se (HPV, ≥CIN II+)=92.5% sp (HPV, ≥CIN II+)=61.8%	Quality Score=7 Ref. Std.: 2 Blind: 0 Verification: 2 Consecutive: 0 Spectrum: 1 Publication: 1 Industry: 1

Chapter I. Introduction

Figure 1. Map of Cervical Cytology Classification Schemes

Classification System	Within Normal Limits	Benign Cellular Changes	Epithelial Cell Abnormalities					
				Squamous Intraepithelial Lesion (SIL)				Invasive Carcinoma
				Low Grade (LSIL)		High Grade (HSIL)		
					Cervical Intraepithelial Neoplasia (CIN)			
					Grade 1	Grade 2	Grade 3	
The Bethesda System[28]	Normal	Infection Reactive Repair	ASCUS*	Condyloma	Mild Dysplasia	Moderate Dysplasia	Severe Dysplasia	In situ Carcinoma
Richart[29]				– – –				
Reagan[30] WHO			Atypia					
Nyirjesy[31]	I		II		III		IV	V

Source: McCrory et al., 1999[32]
* ASCUS, Atypical squamous cells of uncertain significance.

Chapter 1. Introduction

Figure 2. Screening for Cervical Cancer: Analytic Framework

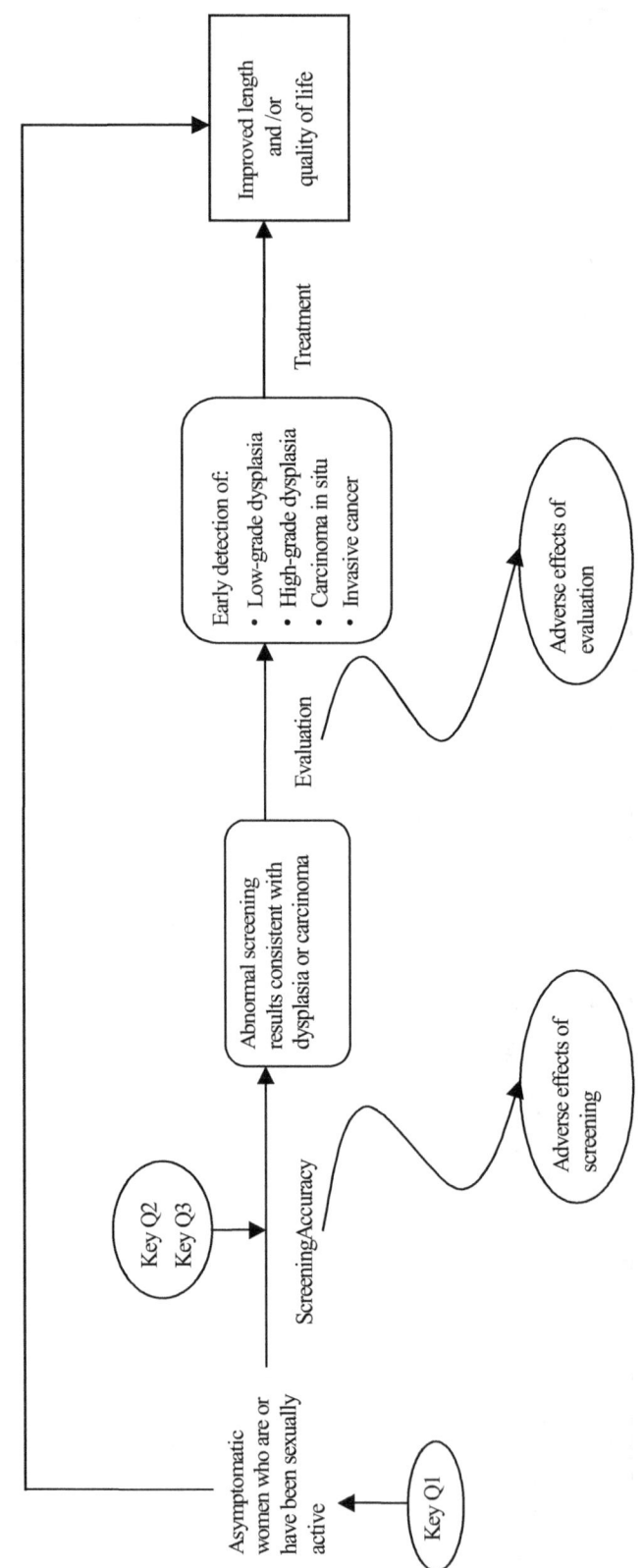

Key Questions
KQ1 Who should be screened for cervical cancer and how often?
KQ2 To what extent do new methods for preparing or evaluating cervical cytology improve diagnostic yield compared to conventional methods? At what cost (harms and economic)?
KQ3 What is the role of HPV testing in cervical cancer screening strategies?

Chapter I. Introduction

Table 1. Technical Characteristics of HPV Testing Methods

Test Technique	Low Sensitivity and Specificity	High Sensitivity and Specificity	Complex Execution/Low Potential for Automation	Suitable for High Throughput/ Amenable to Automation
Southern blot	√		√	
In situ hybridisation protocols*	√		√	
Dot blot procedures	√			
Hybrid Capture I	√		√	
Hybrid Capture II		√		√
PCR		√		
Consensus PCR†		√		√

Source: Adapted from Cuzick et al., 1999.[21]
* Includes Flourescent In Situ Hybridisation (FISH).
†PCR, polymerase chain reaction.

Chapter I: Introduction

Table 2: Recommendations of Other Groups about Pap Smear for Cervical Cancer Screening

Organization	Test Parameters	Screening Interval	Definition of High Risk	Interval for High Risk	Age to Stop Screening
American College of Obstetricians and Gynecologists, 1995	Onset of sexual activity or age 18, whichever occurs first	Annual; after 3 years of normal Paps interval may be lengthened	Women with multiple sexual partners or whose male sexual partners have had multiple partners; sexual intercourse at an early age; women whose male sexual partners have had other sexual partners with cervical cancer; women with current or prior HPV or condylomata or both; women with current or prior herpes simplex virus infections; women infected with HIV; women with a history of STDs; women who are immunosuppressed; smokers and abusers of other substances, including alcohol; women who have a history of cervical dysplasia or cervical cancer or endometrial, vaginal, or vulvar cancer; women of lower socioeconomic status	At physician discretion	No end date

Chapter I: Introduction

Table 2: Recommendations of Other Groups about Pap Smear for Cervical Cancer Screening (cont'd)

Organization	Test Parameters	Screening Interval	Definition of High Risk	Interval for "High Risk"	Age to stop screening
American Society of Clinical Pathologists, 1999	Age 18 or sexually active	Once a year	Cancer, precancerous lesions, a variety of infectious conditions	3 - 6 months or colposcopy	Continuing for the rest of her life
American College of Preventive Medicine, Practice Guidelines Committee, 1996	At onset of sexual activity or age 18 if sexual history is unknown	At least 2 initial screening tests 1 year apart; then interval lengthened at discretion of patient and doctor, but not to exceed >3 year interval	Not mentioned	Not mentioned	Age 65, if no abnormal smears in the prior 9 years, unless patient has not been screened adequately

Chapter I: Introduction

Table 2: Recommendations of Other Groups about Pap Smear for Cervical Cancer Screening (cont'd)

Organization	Test Parameters	Screening Interval	Definition of High Risk	Interval for "High Risk"	Age to stop screening
American Academy of Family Physicians, 1996	Women who have ever had sex and have a cervix	At least every 3 years	Not mentioned	Not mentioned	Not mentioned
Canadian Task Force on Preventive Health Care, 1992	Women > age 18 who have had sexual intercourse	Two annual screens, then every 3 years	Early onset of sexual intercourse; many sexual partners; sexual partner with many sexual partners	More frequently than 3 years	Until age 69
American Cancer Society, 2000	Sexually active or Age18 (as prior entries)	Annually until 3 or more consecutive satisfactory examinations, then at physician discretion	Not mentioned	Not mentioned	General cancer check-up recommendations suggest annual exam continue past menopause
Institute for Clinical Systems Improvement, 2000	Sexually active women younger than age 18 and all women aged 18-64. Omit women who have had a hysterectomy with no residual cervix	3 consecutive normal smears and no dysplasia within 5 years, then less frequently, but at least every 3 years	Mandelblatt Risk Factor Table of Relative Risk	Annually until no longer show dysplasia within 5 years	Age 65
UK National Health Service Cancer Screening Progammes, 1999	Age 20. Omit women who have had a total hysterectomy for nonmalignant reasons	At least every 5 years (free cervical smear test)	Certain types of HVP; women with many sexual partners or whose partner has had many partners; long-term use of the pill; women who smoke	Not mentioned	Age 64

Chapter I: Introduction

Table 2: Recommendations of Other Groups about Pap Smear for Cervical Cancer Screening (cont'd)

Organization	Test Parameters	Screening Interval	Definition of High Risk	Interval for "High Risk"	Age to stop screening
Australian National Cervical Screening Program, 1998	Women who have ever been sexually active beginning at age 18-20	Every 2 years	Not mentioned	Not mentioned	Age 70 with 2 normal Paps in prior 5 years
New Zealand National Cervical Screening Programme	Women who have ever had intercourse beginning at age 20. Women who have had a total hysterectomy for a benign condition do not require further screening	Every 3 years, except if > 5 years since last or if first smear, then repeat in 1 year's time	Sexual behavior, smoking, hormonal or contraceptive use Immunocompromised women	More frequent screening is not recommended for women possibly at higher risk of cervical cancer as there is no evidence that such women have a shorter duration of the preinvasive stage. Immunocompromised women should be screened annually.	Age 70

Chapter II: Methods

Table 3. Overall Inclusion and Exclusion Criteria

Element	Inclusion	Exclusion
Databases	MEDLINE	Other databases
Languages	English	Other languages
Populations	Human	Animal studies
Study Design	Primary research reports, cost-effectiveness analyses, meta-analyses, systematic reviews	Letters, editorials (i.e., no original data)

Chapter II: Methods

Table 4.   Literature Search Results (1995-2000)

Step	Search History	Results
1	Explode cervical neoplasm	29,318
2	Explode cervical dysplasia	2,331
3	Explode mass screening	44,349
4	Explode vaginal smears and screening	2,302
5	1 or 2	29,913
6	3 or 4	45,324
7	5 and 6	3,256
8	Limit 7 to human and English language and year = 1995-2000	**962**

Chapter II: Methods

Table 5. Disposition of Articles Identified by Literature Search

Categorization by Abstract Review	Total Identified	Number Excluded	Retained for Background	Retained for Review
Key Question 1: Who should be screened?	351	205	28	118
Key Question 2: New cytology methods for screening	196	143	5	48
Key Question 3: What is the role of HPV testing?	64	31	3	30
Commentary/opinion, guidelines, methodologic critique, reviews	128	96	32	----
Methods to promote uptake and continuance of screening or improve follow-up of abnormal results	223	216	7	----
Total	962	691	75	196

Chapter III: Results

Table 6. Performance of ThinPrep® in a Prospective Cohort*

ThinPrep Cytology Threshold	Final Diagnosis	Number with Diagnosis (%)	"Equivocal"†	Actual Sens‡ (%)	Estimated Spec§ (%)	PPV‖ (%)	NPV¶ (%)	+LR#	-LR**
≥ ASCUS††	≥ LSIL ‡‡	323 (3.7)	**With Normal**	87.9	**90.5**	95.9	99.5	9.3	0.13
≥ ASCUS	≥ LSIL	1019 (11.8)	With LSIL	55.4	93.0	51.6	94.0	7.9	0.48
≥ LSIL	≥ LSIL	323 (3.7)	**With Normal**	79.6	97.7	57.8	99.2	34.6	0.21
≥ LSIL	≥ LSIL	1019 (11.8)	With LSIL	42.7	99.9	97.8	92.9	42.7	0.57
≥ HSIL§§	≥ HSIL	137 (1.6)	**With < HSIL**	67.2	99.3	61.3	99.5	98.4	0.33
≥ LSIL	≥ HSIL	137 (1.6)	**With < HSIL**	89.8	96.1	25.8	99.8	23.1	0.11

*Data summarized from Hutchinson et al., 1999.[51]
†Category in which final case diagnoses of "equivocal" were assigned for the calculation of test characteristics of ThinPrep; 696 of 8,636 subjects (8.1%) had a final diagnosis of equivocal "conferred...with various combinations of results, including a single cytologic diagnosis of LSIL by any method, and isolated positive cervigram, or equivocal results based on the review of all available data."[51] (50)
‡ Sens, sensitivity
§ Spec, specificity
‖ PPV, positive predictive value
¶ NPV, negative predictive value
# +LR, positive likelihood ratio;
** -LR, negative likelihood ratio.
†† ASCUS, atypical squamous cells of uncertain significance
‡‡ LSIL, low-grade squamous intraepithelial lesion
§§ HSIL, high-grade squamous intraepithelial lesion

Chapter III: Results

Table 7.   Studies with Screening Uses of HPV* Testing

Article	Population	Prevalence of LSIL[†] (%)	Prevalence of HSIL[‡] (%)	Prevalence of Carcinoma (%)
Cuzick et al., 1999[21]	2,988 women age ≥ 34 presenting for routine Pap screening in UK	HPV/borderline = 1.9; CIN[§] 1 = 0.9	CIN[§] 2 = 0.3; CIN[§] 3 = 1.1	No cases
Petry et al., 1999[61]	138 HIV infected women	CIN[§] 1 = 14.5	CIN[§] 2 = 3.6; CIN[§] 3 = 6.5	Invasive Ca[‖] = 2.2
Schiffman et al., 2000[65]	8,554 Costa Rican women from random door-to-door selection	Equivocal = 7.7; LSIL = 2.2	HSIL = 1.5	Invasive Ca[‖] = 0.14
Womack et al., 2000[63]	466 women age 25-55 in primary care clinics in Zimbabwe	LSIL = 12.9	HSIL = 12.0	No cases
Womack et al., 2000[62]	2,140 women age 25-55 in primary care clinics in Zimbabwe	LSIL = 16.2	HSIL = 10.0	Invasive Ca[‖] = 0.14
Wright et al., 2000[64]	1,365 unscreened women, age ≥ 35 from clinics in South Africa	LSIL = 2.9	HSIL = 3.4	Invasive Ca[‖] = 0.7

* HPV, human papilloma virus.
† LSIL, low-grade squamous intraepithelial lesion.
‡ HSIL, high-grade squamous intraepithelial lesion.
§ CIN, cervical intraepithelial neoplasia.
‖ Invasive Ca, invasive cancer.

Chapter III: Results

Table 8. Performance of Screening HPV* Testing for Detection of High-grade Abnormalities

Article	Test Method	Total N	True Positive	False Neg	Sens[†]	True Neg	False Positive	Spec[‡]	Prev[§]	PPV[‖]	NPV[‖]	Pos LR	Neg LR
Cuzick et al., 1999[21]	SHARP[#]	2988	31	11	73.8	2801	145	95.1	1.4	17.6	99.6	15.0	0.3
	HC I**	1285	11	5	68.8	1024	245	80.7	1.3	4.3	99.5	3.6	0.4
	HC II	1703	20	1	95.2	1586	96	94.3	1.2	17.2	99.9	16.7	0.1
Petry et al., 1999[61]	HC I (HIV+[††])	138	16	1	94.1	97	24	80.2	12.3	40.0	99.0	4.7	0.1
Shiffman et al., 2000[65]	HC I	8554	104	35	74.8	7859	556	93.4	1.6	15.8	99.6	11.3	0.3
	HC II	1119	122	16	88.4	873	108	89.0	12.3	53.0	98.2	8.0	0.1
Womack et al., 2000[63]	HC II (HIV+[††])	249	39	4	90.7	85	121	41.3	17.3	24.4	95.5	1.5	0.2
	HC II** (HIV−[††])	217	8	5	61.5	152	52	74.5	6.0	13.3	96.8	2.4	0.5
Womack et al., 2000[62]	HC II	2140	174	41	80.9	1185	740	61.6	10.0	19.0	96.7	2.1	0.3
Wright et al., 2000[64]	HC II	1365	47	9	83.9	1081	228	82.6	4.1	17.1	99.2	4.8	0.2
Total all HC II		6793	410	76	84.4	4962	1345	78.7	7.2	23.4	98.5	4.0	0.2
HCII excluding HIV +		6544	371	72	83.7	4877	1224	79.9	6.8	23.3	98.5	4.2	0.2
HCII in populations with prevalence <10%		5425	249	56	81.6	4004	1116	78.2	5.6	18.2	98.6	3.8	0.3

*HPV, human papilloma virus.
[†]Sens, sensitivity.
[‡]Spec, specificity.
[§]Prev, prevalence.
[‖]PPV and NPV, positive and negative predictive value.
[¶]Pos LR and Neg LR, positive and negative likelihood ratio.
[#]SHARP, SHARP detection system.
**HC, hybrid capture.
[††]HIV, human immunodeficiency virus, where "+" indicates participants with HIV infection and "should be" indicates those without infection.
[‡‡]Physician collected samples.

Chapter III. Results

Table 9. Performance of Screening HPV* Testing for Detection of Low-grade or More Severe Abnormalities

Article	Test Method	Total N	True Pos.	False Neg.	Sens†	True Neg.	False Pos.	Spec‡	Prev§	PPV‖	NPV‖	Pos. LR	Neg. LR
Cuzick et al. 1999[21]	SHARP#	2988	81	45	64.3	2767	95	96.7	4.22	46.0	98.4	19.4	0.43
	HC I**	1285	29	18	61.7	1011	227	81.7	3.66	11.3	98.3	3.4	0.50
	HC II	1703	36	23	61.0	1564	80	95.1	3.46	31.0	98.6	12.5	0.43
Womack et al. 2000[63]	HC II (HIV+††)	249	73	12	85.9	77	87	47.0	34.1	45.6	86.5	1.62	0.30
	HC II (HIV-††)	217	14	17	45.2	140	46	75.3	14.3	23.3	89.2	1.83	0.73
Womack et al. 2000[62]	HCII	2140	359	202	64.0	1024	555	64.9	26.2	39.3	83.5	1.82	0.56
Wright et al. 2000[64]	HC II	1365	78	18	81.3	1072	197	84.5	7.6	28.4	98.3	5.42	0.22
Total all HC II		5674	560	272	67.3	3877	965	80.1	14.7	36.7	93.4	3.4	0.40
HC II excluding HIV+		5425	487	260	65.2	3800	878	81.2	13.8	35.7	93.6	3.5	0.40
HC II in populations with prevalence <10%		3068	114	41	73.5	2636	277	90.5	5.1	29.2	98.5	7.7	0.30

*HPV, human papilloma virus.
§Prev, prevalence.
‖PPV and NPV, positive and negative predictive value.
†Sens, sensitivity.
¶Pos LR and Neg LR, positive and negative likelihood ratio.
‡‡Physician collected samples.
‡Spec, specificity.
#SHARP, SHARP detection system.
**HC, hybrid capture.
††HIV, human immunodeficiency virus, where "+" indicates participants with HIV infection and "should be" indicates those without infection.

Chapter III. Results

Table 10. HPV Testing Among Women with Abnormal Pap Test Results*

Article	Population/ Referral Pap	Prevalence of LSIL (%)	Prevalence of HSIL (%)	Prevalence of Carcinoma (%)
Adam et al., 1998[71]	454 indigent Texan women referred for colposcopy after ASCUS or LSIL	LSIL = 55.3	HGSIL = 14.5	4 cases invasive cancer = 0.40
Bollen et al., 1997[66]	190 consecutive Dutch women with mild or moderate dysplasia	LSIL = 57.8	HSIL = 29.5	No cases
Herrington et al., 1995[69]	167 British women referred for colposcopy for low grade cytologic findings (≤ CIN 1)	"wart virus changes" = 37.7 CIN 1 = 15.0	CIN 2 = 7.2 CIN 3 = 16.8	No cases
Hillemanns et al., 1999[72]	247 German colposcopy patients	CIN 1 = 7.3	CIN 2/3 = 15.4	2 cases invasive cancer = 0.81
Manos et al. 1999[68]	973 US HMO patients with ASCUS	LSIL = 12.8	HSIL = 6.7	1 case invasive cancer = 0.1
Sigurdsson et al., 1997[67]	358 Icelandic women referred for colposcopy	CIN 1 = 16.5	CIN 2 = 15.9 CIN 3 = 36.6	7 cases invasive cancer = 2.0
Sun et al., 1995[73]	520 US or Canadian women referred for colposcopy	LSIL = 31.0	HSIL = 18.8	7 cases invasive cancer = 1.3

*HPV, human papilloma virus; LSIL, low-grade squamous intraepithelial lesion; HSIL, high-grade squamous intraepithelial lesion; CIN, cervical intraepithelial neoplasia

Table 11. HPV Testing as a Triage Tool Among Women with an Abnormal Pap Test for Detection of HSIL*

Author (Year)	Test Method	Total N	True Pos.	False Neg	Sens	True Neg	False Pos.	Spec	Prev HGSIL	PPV	NPV	Pos LR	Neg LR
Adam et al., 1998[71]	Consensus PCR	454	43	23	65.2	233	155	60.1	14.5	21.7	91.0	1.6	0.6
Bollen et al., 1997[66]	SHARP PCR	190	53	3	94.6	54	80	40.3	29.5	39.8	94.7	1.6	0.1
Herrington et al., 1995[69]	Consensus PCR	167	35	5	87.5	79	48	62.2	24.0	42.2	94.0	2.3	0.2
Hillemanns et al., 1999[72]	HC II	247	35	3	92.1	150	59	71.8	15.4	37.2	98.0	3.3	0.1
Manos et al., 1999[68]	HC II	973	58	7	89.2	582	326	64.1	6.7	15.1	98.8	2.5	0.2
Sigurdsson et al., 1997[67]	Consensus PCR	358	156	39	80.0	118	45	72.4	54.5	77.6	75.2	2.9	0.3
	HC I	358	136	59	69.7	96	67	58.9	54.5	67.0	61.9	1.7	0.5
Sun et al., 1995[73]	Consensus PCR	520	87	18	82.9	145	270	34.9	20.2	24.4	89.0	1.3	0.5
	HC II	520	83	22	79.0	181	234	43.6	20.2	26.2	89.2	1.4	0.5
Total all PCR		1689	374	88	81.0	629	598	51.3	27.4	38.5	87.7	1.7	0.4
Total all HC II		1740	176	32	84.6	913	619	59.6	12.0	22.1	96.6	2.1	0.3

*HPV, human papilloma virus; HSIL, high-grade squamous intraepithelial lesion.